Transplutonium Elements— Production and Recovery

Transplutonium Elements— Production and Recovery

James D. Navratil, EDITOR
International Atomic Energy Agency

Wallace W. Schulz, EDITOR
Rockwell Hanford Operations

Based on a symposium

cosponsored by the Divisions of

Industrial and Engineering Chemistry

and Nuclear Chemistry and Technology

at the Second Chemical Congress

of the North American Continent

(180th ACS National Meeting),

Las Vegas, Nevada,

August 27–28, 1980.

A C S   S Y M P O S I U M   S E R I E S **161**

AMERICAN CHEMICAL SOCIETY
WASHINGTON, D. C.   1981

Library of Congress CIP Data

Transplutonium elements, production and recovery.
(ACS symposium series, ISSN 0097-6156; 161)

"Based on a symposium cosponsored by the Divisions
of Industrial and Engineering Chemistry and Nuclear
Chemistry and Technology at the Second Chemical
Congress of the North American Continent (180th
ACS national meeting), Las Vegas, Nevada, August
27–28, 1980."

Includes bibliographies and index.

1. Transplutonium elements—Congresses.
I. Navratil, James D., 1941–   . II. Schulz, Wallace W. III. Chemical Congress of the North American Continent (2nd: 1980: Las Vegas). IV. American
Chemical Society. Division of Industrial and Engineering Chemistry. V. American Chemical Society. Division of Nuclear Chemistry and Technology. VI. Title.
VII. Series: ACS symposium series; 161.

QD172.T65T7   621.48'335   81-7999
ISBN 0–8412–0638–4 AACR2 ACSMC8 161 1–302
                                            1981

ACS Symposium Series

M. Joan Comstock, *Series Editor*

FOREWORD

The ACS SYMPOSIUM SERIES was founded in 1974 to provide a medium for publishing symposia quickly in book form. The format of the Series parallels that of the continuing ADVANCES IN CHEMISTRY SERIES except that in order to save time the papers are not typeset but are reproduced as they are submitted by the authors in camera-ready form. Papers are reviewed under the supervision of the Editors with the assistance of the Series Advisory Board and are selected to maintain the integrity of the symposia; however, verbatim reproductions of previously published papers are not accepted. Both reviews and reports of research are acceptable since symposia may embrace both types of presentation.

CONTENTS

PREFACE

Certain of the transplutonium elements are used extensively in nuclear gauges and in many other fields as well. Industrial-scale production of these man-made elements requires development and application of appropriate recovery, separation, and purification processes.

The 17 papers in this volume provide authoritative, in-depth coverage of an important area of nuclear and industrial chemistry. In addition to 12 U.S. authored papers, there are papers from authors in France, Japan, Peoples Republic of China, Sweden, and West Germany. This volume thus includes contributions from most countries in the world that have significant transplutonium element production and recovery programs and facilities.

We believe that this collection of papers will provide members of the nuclear community and chemists and engineers everywhere a comprehensive review of what is currently going on at the "bottom of the Periodic Table."

JAMES D. NAVRATIL
Vienna, Austria

WALLACE W. SCHULZ
Richland, Washington

January 1, 1981

INTRODUCTION

This collection of the state-of-the-art papers emphasizes the continuing importance of industrial-scale production, separation, and recovery of transplutonium elements. Americium (At. No. 95) and curium (At. No. 96) were first isolated in weighable amounts during and immediately after World War II. Berkelium and californium were isolated in 1958 and einsteinium in 1961. These five man-made elements, in each case, subsequently became available in increasing quantities.

The U.S. transplutonium element production programs in the 1940s, 1950s, and early 1960s used, successively, nuclear reactors at Oak Ridge, Hanford, Chalk River (Canada), and Idaho. Higher-flux reactors at Savannah River and Oak Ridge were used in the late 1960s and during the 1970s for production of kilogram amounts of both americium and curium, grams of californium, and milligrams of berkelium and einsteinium. The transeinsteinium elements up through element 106 are produced, by bombardment of lighter actinide isotopes with heavy ions, in tracer quantitites that continue to diminish, and very much so, with atomic number.

GLENN T. SEABORG
Berkeley, California

PRODUCTION OF
TRANSPLUTONIUM ELEMENTS

Production of Transplutonium Elements in the High Flux Isotope Reactor

J. E. BIGELOW, B. L. CORBETT, L. J. KING,
S. C. McGUIRE, and T. M. SIMS

Oak Ridge National Laboratory, P.O. Box X, Oak Ridge, TN 37830

The National Transplutonium Element Production Program was established in the late 1950's to concentrate the "large-scale" production of transplutonium elements at a central location. These products are then distributed to researchers throughout the country upon the recommendations of a Transplutonium Program Committee which is comprised of representatives from the major laboratories which have an interest in transplutonium element research. The Oak Ridge National Laboratory was selected as the site for these production facilities, consisting of a high flux reactor and an adjacent radiochemical processing plant, which are capable of producing gram amounts of ^{252}Cf and related quantities of the other heavy elements (1). These man-made elements are all intensely radioactive and can be processed safely and reliably only in an elaborate remote handling facility, such as the Transuranium Processing Plant (TRU). This facility and some of the processes carried out therein for recovery and purification of transplutonium elements are described in other papers in this symposium (2,3,4,5). We have now made over 1000 shipments of these products to 30 different laboratories throughout the U.S. and in several foreign countries, attesting to the success of the Program.

All of this would not be possible without the High Flux Isotope Reactor (HFIR) to serve as a source of neutrons to carry out the transmutation of the elements. Since first reaching full power (100 MW) on October 21, 1966, the HFIR has logged 4148 equivalent full power days through December 31, 1979, for an overall operating efficiency of 86%. During many years, this figure has run 93% or more.

The purpose of this paper is to indicate the capabilities of the HFIR for transplutonium element production and particularly to dwell on the mathematical techniques involved in forecasting the composition of irradiated target materials. Also described are some of the uses to which such forecasts are put. Early work along this line was published by Burch, Arnold, and Chetham-Strode (6), providing the basis for design of HFIR and TRU.

0097-6156/81/0161-0003$05.00/0
© 1981 American Chemical Society

Transmutation Reactions

Figure 1 is a portion of the chart of the nuclides which includes those nuclides which are formed by neutron irradiation and decay from ^{242}Pu, our original starting material. When a nucleus captures a neutron, the mass number increases by 1 and the new nucleus will be represented by the square to the right. This process will continue producing heavier and heavier isotopes until a nuclide is formed that has a high probability of decaying before it can react with another neutron. If the decay is a beta decay, a new element is formed (this is represented by a move diagonally upward to the left). Continued irradiation produces isotopes of this new element until another beta decay produces still another element. The process terminates at ^{258}Fm because that isotope decays by spotaneous fission with a half-life of 380 μs and no beta-active isotope of fermium is formed before this point is reached. Other natural decay processes can occur besides beta decay, such as alpha decay, electron capture, and isomeric transition. Neutron-induced processes, besides capture, include fission and various spallation reactions. With a couple of minor exceptions, the latter are not very important in the transplutonium element region and they are not modeled in our calculations.

The High Flux Isotope Reactor

The High Flux Isotope Reactor (HFIR) was designed to produce very intense neutron fluxes (>10^{19} neut$\cdot$m$^{-2}\cdot$s^{-1}) expressly for the production of transplutonium elements (7). The various core components are arranged in concentric cylindrical regions, all of which have a height of about 0.6 m. The innermost region is a flux trap containing the target island. This is surrounded by the two-piece annular fuel assembly. A new fuel assembly contains initially 9.4 kg of ^{235}U and can operate 21-23 days before replacement. Farther out from the centerline are the control cylinders, inner and outer, and lastly the beryllium reflector which is made up of several annular segments to facilitate replacement as required by radiation damage. The beryllium reflector region is penetrated by a number of thimbles which are very useful for isotope production or irradiation experiments of many kinds (8).

The primary coolant (light water) is admitted to the pressure vessel at 49°C and at a pressure of 5.2 MPa. The coolant flow of 1 m^3/s results in a temperature rise of 24°C and a pressure drop of 0.76 MPa as the coolant flows through the reactor, removing the 100 MW of fission heat.

The target island contains 31 positions for the aluminum-clad target assemblies. As presently operated, the target in the centerline position is replaced with a versatile hydraulic rabbit facility, which gives ready access to the position of

highest flux in the reactor. From time to time, other target
positions have been replaced by special experimental assemblies, so that the number of target positions available for
transplutonium element production fluctuates between 27 and 30.

A target assembly is shown in Fig. 2 in a cutaway view to
show the interior features. The actinide oxide--aluminum powder
blend is pressed into pellets, 35 of which are loaded into an
aluminum tube fitted with aluminum alloy liners to maintain a
void at each end. Plugs are welded into the ends to encapsulate
the radioactive material. The upper end plug also serves as a
remote handling fixture. A coolant flow shroud is mechanically
attached to the outside of the tube providing each target rod
with its own coolant channel, as well as maintaining the hexago-
nal lattice spacing in the reactor. Target assemblies may be
loaded with up to 10 g of ^{242}Pu, ^{243}Am, or ^{244}Cm, or any combination of the above, including equilibrium amounts of heavier
isotopes, for a total of 10 g of heavy metal (11.15 g of oxide).
The ^{242}Pu targets for the initial reactor loading were fabricated in a glove box facility (9), but the other materials are
all sufficiently radioactive as to require remote fabrication in
the TRU hot cells (10).

Model of Flux in the HFIR Target Island

The neutron flux (the product of neutron concentration and
velocity) is a strong function of neutron energy, position in
the target island, and the reactor operating conditions. The
means by which these variations are handled is discussed below.

A 2-group set of fluxes is used for estimating the transmutation of actinide elements. All neutrons having an energy less
than 39.9 kJ/mol (0.414 eV) are considered "thermal." All
neutrons having energies between 39.9 kJ/mol and 9.75 MJ/mol
(101 eV) were tallied and the result divided by 5.5, the number
of lethargy units spanned by this energy range. (Lethargy is
related to the negative logarithm of the energy.) This latter
flux is called the "resonance" or "epithermal" flux per unit
lethargy. The values of these two flux groups calculated for
the original reactor neutronic design are shown on Fig. 3 as a
function of radial distance from the reactor centerline for
several different assumed target loadings. The vertical lines
represent the radial positions of the various groups of target
assemblies. One particular target loading was chosen to represent the loadings typically encountered in regular operation and
the intersection of those curves (for both groups) with the
various target positions were then designated as Standard
Midplane Fluxes for that ring of targets.

The axial distribution was measured in early experiments in
the HFIR. The data were very well fit by the usual chopped
cosine distribution with a small amount of reflector peaking
(Fig. 4). We generally calculate the target compositions at

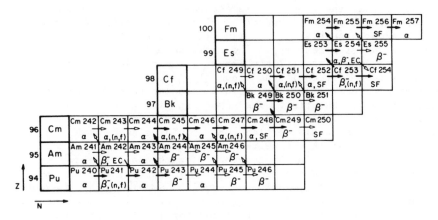

Figure 1.   Transuranium nuclide production paths

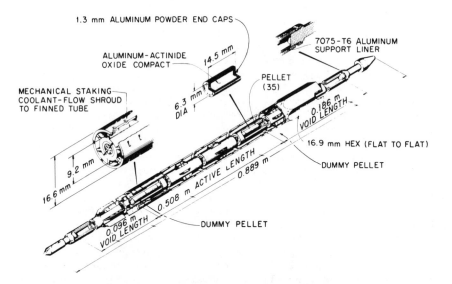

Figure 2.   Diagram of HFIR target assembly

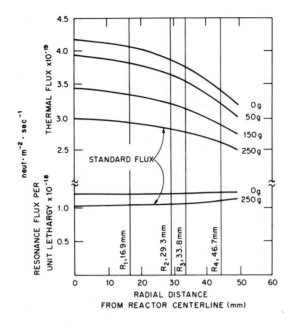

Figure 3. Radial flux distribution in HFIR target island

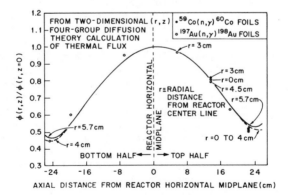

Figure 4. Axial flux distribution in HFIR target island

various places along the assembly using the flux appropriate to
that location and then axially average by Simpson's Rule.

In most reactors operating at a constant power level, the
flux increases with time as the fuel is consumed. This is not
true within the HFIR target island (because of its flux-trap
design) where the neutron fluxes are essentially constant
throughout the length of an operating cycle (about 23 days).
Therefore, a time-average flux can be calculated which is proportional to the reactor thermal power. When calculating exact
histories of target assemblies, the power data are taken from
the reactor operating logs. For design studies, a constant flux
for a 23-day period is assumed.

Another potential variation in the thermal flux, i.e. local
perturbations, is assumed to be negligible because of the relatively small quantities of transplutonium elements contained in
an individual target assembly.

Cross-Section Model

Neutron cross sections are a measure of the probability of
neutrons interacting with a given nucleus. The rate at which a
given reaction occurs is given by the product of the number of
atoms of the nuclide, N, its microscopic cross section, σ, and
the neutron flux, ϕ. Since different kinds of interactions are
possible (e.g., neutron capture, scattering, fission), a cross
section is associated with each of these processes and the
various cross sections are additive. The cross sections are a
very strong function of the incident energy of the neutron, and
some means of folding this information into the spectrum of
neutrons must be utilized. Fortunately, as far as computing
transplutonium element production in the HFIR is concerned, we
need only consider interactions in the thermal and epithermal
energy regions. The two regions are modeled differently.

Thermal Cross Section. In most of the nuclides of
interest, the cross section in the thermal region varies inversely with the neutron velocity, v, which is proportional to the
square root of the neutron kinetic energy. The neutron energy
spectrum in this same region is reasonably well approximated by
a Maxwell-Boltzman distribution in thermal equilibrium with the
light-water moderator (which is estimated to average about 54°C
in the flux trap). Two conventions are used here: (1) the
thermal cross section used in our calculations is the cross section for interaction with neutrons having a velocity of 2200
m/s, which corresponds to an energy of 2.41 kJ/mol (0.025 eV),
the most probable energy for neutrons in thermal equilibrium at
293.15 K. The symbol for this cross section is σ_{2200}; (2) the
thermal flux (ϕ_{th}) is represented by an "equivalent 2200 m/s"
flux (ϕ_{2200}) which is that flux which, when multiplied by σ_{2200},
yields the same reaction rate as does the actual flux multiplied

by the actual cross sections, when integrated across the entire energy spectrum from 0 to 39.9 kJ/mol. If the cross section truly varied as 1/v and if the neutron energy spectrum were truly in equilibrium at some temperature, T, the ratio between this equivalent flux and the actual flux would be:

$$\frac{\phi_{2200}}{\phi_{th}} = \frac{\sqrt{\pi}}{2} \sqrt{\frac{293.15}{T}}$$

For the HFIR, this ratio is 0.839, and for calculational purposes, the thermal fluxes described in the preceding section were multiplied by this quantity before use in computing reaction rates.

Epithermal Cross Section. In the region of neutron energies slightly higher than thermal (epithermal) the modeling of cross sections is quite different. In this region, there are sharp peaks in the cross sections at certain energies where the kinetic energy plus the binding energy of the neutron in the nucleus matches an energy state of the compound nucleus. This phenomenon is called resonance absorption; thus, this energy region is frequently referred to as the "resonance" region. The great amount of detail in the energy-dependent neutron cross-section data (11) makes the evaluation of the overall reaction rate extremely difficult unless some kind of overall averaging can be accomplished. Fortunately, the energy dependence of the flux in this part of the spectrum approaches an idealized case which is exactly the form required to simplify the analysis. So the cross sections can be integrated through the resonance region without involving the fluxes, and then the flux can be included later in the form of the flux per unit lethargy. Thus, the overall reaction rate constant, k, for resonance neutrons will be

$$k = RI \cdot \phi_{res}$$

where RI is the resonance integral of the cross sections and ϕ_{res} is the resonance flux per unit lethargy.

As indicated above, in evaluating ϕ_{res}, the averaging was performed only between 39.9 kJ/mol and 9.75 MJ/mol since the majority of the interactions involving transplutonium nuclides occur in this interval; further, it was a sub-grouping readily available to us from the complete reactor neutronic analysis.

Resonance Self-Shielding. At the energy corresponding to the peak of a given resonance, the absorption cross section can be enormous. Here, the nucleus becomes effectively a sponge, soaking up the vast majority of the neutrons with energies near that of the resonance. In a target of finite thickness (such as the 5-mm-diameter active region in the HFIR target assemblies),

the atoms in the outer layers of the target react strongly with
the incoming neutrons and prevent the neutrons from reaching the
atoms in the interior. This phenomenon is known as resonance
self-shielding, and is a function of the atom density of the
absorbing nuclei, the geometry of the region containing the
absorbing nuclei, and the scattering properties of all nuclides
contained within that region. The relationship below is valid
for a single resonance absorption peak, but for a real nuclide
possessing a multitude of resonances, it should be regarded more
as an empirical correction for resonance self-shielding:

$$RI_{eff} = \frac{RI}{\sqrt{1+CN}}$$

where RI_{eff} is the effective resonance integral, N is the
number of grams of the particular nuclide in one target rod, and
C is a constant incorporating the conversion factors of N into
atom density, as well as the information relating to the target
geometry and neutron scattering properties.

All of these models must now be combined to yield a useful
approximation for the reaction rate of a nuclide with the
neutrons in the HFIR.

$$\text{Reaction Rate} = N(\phi\sigma)_{eff} = N\phi_{2200}\sigma_{2200} + N\phi_{res}\frac{RI}{\sqrt{1+CN}}$$

The constant C was initially calculated for the nuclide
^{242}Pu based on the first major resonance at 259 kJ/mol (2.68 eV)
(12). For some nuclides, values of C were assumed based on the
peak absorption cross section in the major resonance. Others
were assumed based on proportionality to the resonance integral
(which can be measured empirically without knowing the detailed
energy-dependent spectrum). Then, these assumed values for C
and also σ_{2200} were adjusted by trial and error procedures to
produce reasonable agreement with experimentally determined
tranmutation reactions. Table I shows values presently in use
for the parameters σ_{2200}, C, and RI for both capture and fission
for the transuranic nuclides considered in this program.

Both processes occur simultaneously and each is first
order with respect to the reactant. Thus, the rate of change of
the quantity of nuclide N_i is given by

$$\frac{dN_i}{dt} = -\lambda N_i - N_i (\phi_{2200}\,\sigma_{i,2200}^a + \phi_{res}\frac{RI_i^a}{\sqrt{1+C_iN_i}}) + P$$

where most of the symbols were defined before, and the super-
script, a, refers to the sum of neutron capture and fission
processes. P is the production term and is either of the form:

Table I. Neutron cross section parameters used to compute transmutations in HFIR target irradiations

Nuclide	σ^c_{2200}	c^c	RI^c	σ^f_{2200}	c^f	RI^f
^{238}Pu	560	0	150	16.5	0	25
^{239}Pu	265.7	0	195	742.4	0	324
^{240}Pu	290	0	8453	0.05	0	0
^{241}Pu	360	0	166	1011	0	541
^{242}Pu	19.5	6.20	1280	0	0	0
^{243}Pu	80	0	0	210	0	0
^{244}Pu	1.6	0	0	0	0	0
^{245}Pu	277	0	0	0	0	0
^{243}Am	105	0	1500	0	0	0
^{244}Am	0	0	0	2300	0	0
^{244}Cm	10.0	4.0	650	1.2	4.0	12.5
^{245}Cm	343	2.4	120	1727	2.4	1140
^{246}Cm	1.25	0	121	0	0	0
^{247}Cm	60	0	500	120	0	1060
^{248}Cm	3.56	2.0	170	0	0	0
^{249}Cm	2.8	0	0	50	0	0
^{250}Cm	2	0	0	0	0	0
^{249}Bk	1451	2.4	1240	0	0	0
^{250}Bk	350	0	0	3000	0	0
^{249}Cf	450	1.46	750	1690	5.8	2920
^{250}Cf	1900	20	11600	0	0	0
^{251}Cf	2850	14	1600	3750	14	5400
^{252}Cf	19.8	0	44	32	0	110
^{253}Cf	12.6	0	0	1300	0	0
^{254}Cf	50	0	1650	0	0	0
^{253}Es	345	0	0	0	0	0
^{254}Es	20	0	0	3060	0	0
^{254m}Es	1.26	0	0	1840	0	0
^{255}Es	60	0	0	0	0	0
^{254}Fm	76	0	0	0	0	0
^{255}Fm	26	0	0	100	0	0
^{256}Fm	45	0	0	0	0	0
^{257}Fm	10	0	0	5500	0	0

$$+ \lambda N_{i-1}$$

if the nuclide is formed from the precursor by a decay process or:

$$+ N_{i-1}(\phi_{2200}\sigma^c_{i-1,2200} \quad + \phi_{res} \; \frac{RI^c_{i-1}}{\sqrt{1+C_{i-1}N_{i-1}}})$$

if the nuclide is formed from the precursor by neutron capture.

System of Equations. In a target assembly which may contain 20 or more nuclides in significant concentrations, a very complex system of linear differential equations with (nearly) constant coefficients is required to properly model the transmutation reactions. Various methods could be used to solve this system of equations, but A. R. Jenkins, of the ORNL Mathematics Division, recommended for our particular type of problem that we use the analytic solution to the Bateman Equations and that the cross sections which vary slowly as the composition changes be held constant within any one time step. The cross sections are then re-evaluated for each new time step as required to maintain a realistic modeling. Accordingly, he modified the existing CRUNCH Code (13) to take account of the 2-group, 3-parameter cross sections as described above. When this new program was first implemented in 1964 (on the CDC 1604 computer), a typical run required about 20 minutes. Today, on the IBM 360, Model 91, the same job would run in a few seconds.

Applications of Computer Models

Prediction of Target Compositions. One application of the computer program developed for mathematically modeling the transplutonium element transmutations in a HFIR target is that of predicting the amounts of transplutonium elements which will be available from a given irradiation. This information is then used in the planning of processing campaigns. This is also the mechanism for validating the model by comparing calculated and measured values. If significant discrepancies arise, some new values for parameters can be chosen and the process repeated until the calculated values are acceptably close to the measured ones. Table II shows the comparison between calculated and measured values for a recent campaign to process 13 HFIR targets. The exact irradiation histories were included in the computation of each individual target assembly (with some multiplicities) and the results summed.

It can be seen that agreement between calculated and measured values up through mass 253 is probably within the range of analytical uncertainties.

Qualification of Targets for Irradiation. A second application of the model is the prediction of fission rates (and
hence heat fluxes) for targets being transferred to the reactor
for irradiation. The allowable heat fluxes (14) were selected
to prevent melting of aluminum at the center line of the target
assembly and must not be exceeded at any time during the proposed
irradiation of the targets.

Optimization of Irradiation Times. By far the greatest
usage of the calculational model has been to study the optimization of irradiation times. This is a multi-dimensional problem
of great complexity which has as its motivation the proper utilization of very expensive facilties and a very valuable inventory of intermediate products, mainly the mixtures of curium
isotopes. The problem does not lend itself to a complete solution; however, various simplifying approximations can be applied
to the problem to explore the interacting parameters.

The first simplification is to adopt ^{252}Cf as a yardstick
for productivity. The nuclides past ^{252}Cf are all produced more
or less in proportion to the ^{252}Cf. Also, it is the major source
of penetrating radiation so that many features of the design of
TRU and some of the operating schedules were predicated on the
amounts of ^{252}Cf to be processed. Various attempts have been
made (6 15,16) to develop methods of maximizing the ^{252}Cf production rate, usually assuming an unlimited supply of feed
material of a given composition. In the earlier years, the poor
quality of feed available put greater emphasis on this approach.

Since the Californium-I campaign (17) at the U.S. Department of Energy Savannah River site and the consequent availability of a much better quality of feed, the emphasis is shifting
toward efficient utilization of the finite inventory of curium
feedstocks so as to be able to support the transplutonium element research program on a useful scale for an extended period
of time.

More definitions are necessary to attempt this sort of
optimization: Potential californium is a measure of the
maximum amount of californium that can be produced from a given
batch of feed, taking into account the fact that many atoms
undergo fission along the path from feed to product. The efficiency of a particular irradiation is the amount of californium
produced divided by the amount of potential californium consumed
in the irradiation and subsequent processing. This efficiency
measure takes into consideration the destruction of the ^{252}Cf by
decay and neutron capture and processing losses of all the
nuclides in the chain.

The efficiency defined in this manner varies with cumula-
tive irradiation time as shown on Fig. 5. Initially zero, the
efficiency rises as more of the heavier intermediate nuclides
are formed, thus increasing the production rate of ^{252}Cf.
Further irradiation will cause the efficiency to level off and

Table II. Comparison of Calculated and Measured Isotope Yields
for 13 HFIR Targets Processed During TRU Campaign 59.

(The entries in the table correspond to the date of discharge
from the reactor)

Nuclide	Measured Yield	Calculated Yield	δ %
^{244}Cm	6.9 g	5.9 g	−14.5
^{245}Cm	91.1 mg	78.9 mg	−13.4
^{246}Cm	41.0 g	35.9 g	−12.4
^{247}Cm	1.1 g	1.0 g	−9.1
^{248}Cm	8.2 g	10.2 g	−24.4
^{249}Bk	75.8 mg	60.4 mg	−20.3
^{250}Cf	50.9 mg	37.4 mg	−26.5
^{251}Cf	14.4 mg	13.3 mg	−7.6
^{252}Cf	560.1 mg	508.2 mg	−9.3
^{253}Cf	9.1 mg	6.9 mg	−24.2
^{253}Es	2.9 mg	2.4 mg	−17.2
^{254}Es	12.0 μg	111.1 μg	+825.8

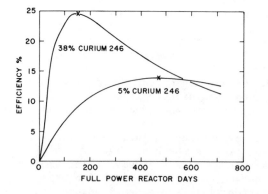

Figure 5. Californium-252 production efficiency for HFIR feed of two compositions

then decline as more and more of the ^{252}Cf already produced is consumed by decay, neutron capture or fission. Efficiency curves for two different isotopic mixtures of curium feed are traced on Fig. 5. The behavior is similar although the better quality curium (higher % ^{246}Cm) produces californium sooner and with higher efficiency. The maximum efficiency points are marked with x's. Figure 6 shows the actual amount of ^{252}Cf produced in a HFIR target assembly loaded with 10 g of each of the two curium compositions. The scale showing irradiation time is the same as for Fig. 5 and the two x's mark the irradiation times which were determined as the respective maximum efficiencies on Fig. 5. Note that there is very little about the shape of the curves on Fig. 6 that would cause one to select those particular points.

With the above defined procedure for optimizing the length of the irradiation time as a function of the quality of curium in the successive recycles, an attempt was made to determine the maximum ^{252}Cf production obtainable in the TRU-HFIR complex. Various additional simplifying assumptions concerning scheduling, target loading, processing losses, and ^{252}Cf decay before shipment, were chosen to facilitate the computations and yet be as realistic as possible.

The results of this exercise are presented as the uppermost curve in Fig. 7. The feed material to the first cycle is curium originating from the Cf-I campaign at Savannah River. Curium residues from the first cycle (and each subsequent cycle) are recycled back to the reactor after a period of time to represent processing and refabrication. Additional first-cycle targets are used to fill the remaining spaces. The numbers in circles represent the annual rate of target processing at various stages along the curve. The numbers increase because the average quality of the curium improves with recycle, thus shortening the optimum irradiation period. Figure 7 also shows the effect of determining the reactor irradiation period by some other technique than the "optimum." In these lower curves, all conditions were the same except the number of targets processed per year is arbitrarily limited to 50 and 25, respectively.

Practical Considerations

In spite of the natural effort to strive toward the "optimum," we are constantly forced to recognize reality. Irradiation times were shorter than optimum in the early days because of problems associated with target failures (18). Later, as these problems were overcome, they also became moot, as the improving quality of feed available caused dramatic reductions in the optimum irradiation time.

Other considerations have a strong bearing on the actual decisions concerning a processing campaign. One wishes to avoid campaigns during prime vacation periods because the reduced

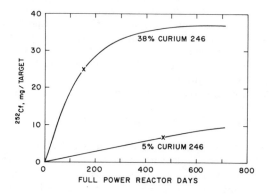

Figure 6.   Californium-252 production for HFIR feed of two compositions

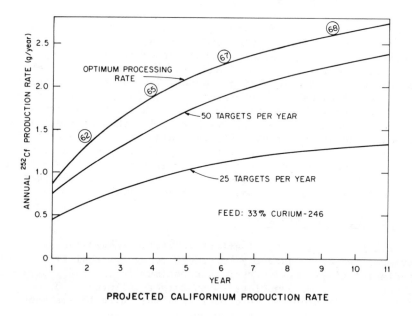

PROJECTED CALIFORNIUM PRODUCTION RATE

Figure 7.   Optimized californium production rates

manpower available slows the processing rate. Schedules must be adjusted to fit maintenance needs. Product transportation is a factor, too, especially for short-lived products. For example, we learned not to ship out ^{257}Fm during the Christmas rush. The timing of the customer's ability to utilize the material sometimes affects decisions, as the use of cyclotrons, lasers, and other such expensive research facilities must be carefully scheduled long in advance. Even the scheduling of international symposia on transplutonium elements may have an impact on us.

Conclusions

The techniques described here have been demonstrated to predict the contents of transplutonium element production targets, at least for isotopes of mass 253 or less. The HFIR irradiation model is a workhorse for planning the TRU processing campaigns, for certifying the heat evolution rate of targets prior to insertion in the reactor, for predicting future production capabilities over a multi-year period, and for making optimization studies. Practical considerations, however, may limit the range of available options so that "optimum" operation is not always achievable. We do intend, however, to keep "fine-tuning" the constants which define the cross sections as time permits. We need to do more work on optimizing the production of ^{250}Cm, ^{254}Es, ^{255}Es, and ultimately ^{257}Fm, since researchers are interested in obtaining larger quantities of these rare and difficult-to-produce nuclides.

Acknowledgements

The authors wish to acknowledge the considerable contributions of Dr. Joseph Halperin, Staff Chemist at ORNL, whose advice and counsel guided the early stages of development of our model of neutron fluxes and cross sections.
This research was sponsored by the Office of Basic Energy Sciences, U. S. Department of Energy, under contract W-7405-eng-26 with the Union Carbide Corporation.

Literature Cited

1. Ferguson, D. E. Nucl. Sci. Engrg. 17, 1963, 435 ff.
2. King, L. J.; Bigelow, J. E.; Collins, E. D. "Industrial-Scale Production - Separation - Recovery of Transplutonium Elements," ACS Symposium, 2nd Chem. Congr. North American Continent, 1980.
3. Collins, E. D.; Benker, D. E.; Chattin, F. R.; Orr, P. B.; Ross, R. G. "Industrial-Scale Production - Separation - Recovery of Transplutonium Elements," ACS Symposium, 2nd Chem. Congr. North American Continent, 1980.

4. Benker, D. E.; Chattin, F. R.; Collins, E. D.; Knauer, J. B.;
    Orr, P. B.; Ross, R. G.; Wiggins, J. T. "Industrial-
    Scale Production – Separation – Recovery of Trans-
    plutonium Elements," ACS Symposium 2nd Chem. Congr. North
    American Continent, 1980.
5. Chattin, F. R.; Benker, D. E.; Lloyd, M. H.; Orr, P. B.;
    Ross, R. G.; Wiggins, J. T. "Industrial-Scale Production –
    Separation – Recovery of Transplutonium Elements," ACS
    Symposium, 2nd Chem. Congr. North American Continent, 1980.
6. Burch, W. D.; Arnold, E. D.; Chetham-Strode, A. Nucl. Sci.
    Engr., 17, 1963, 438.
7. Binford, F. T.; Cramer, E. N. (Editors), The High-Flux Iso-
    tope Reactor – A Functional Description, ORNL-3572 (May 1964).
8. Sims, T. M.; Swanks, J. H. High Flux Isotope Reactor
    (HFIR) Experiment Facilities and Capabilities, ORNL
    Brochure available from K. J. Foust, Bldg. 7910, Oak Ridge
    National Laboratory, P. O. Box X, Oak Ridge, TN 37830.
9. Sease, J. D. The Fabrication of Target Elements for the
    High Flux Isotope Reactor, ORNL-TM-1712 (March 1967).
10. Van Cleve, J. E. Jr.; and Williams, L. C. "Hot Cell
    Fabrication of Target Rods and Neutron Sources," Welding
    Journal, August 1973, 497.
11. Weston, L. W. "Review of Microscopic Neutron Cross
    Section Data for the Higher Plutonium Isotopes in the
    Resonance Region," Proc. Specialists Mtg. on Nucl. Data
    of Higher Plutonium and Americium Isotopes for Reactor
    Applications, Brookhaven National Laboratory, Nov. 20–22,
    1978, BNL-50991, p. 1 (1979).
12. Young, T. E.; and Reeder, S. D. "Total Neutron Cross
    Section of ^{242}Pu," Nucl. Sci. Eng., 40, 384–395 (1970).
13. Lietzke, M. P.; and Claiborne, H. C. "CRUNCH – An IBM-704
    Code for Calculating N Successive First-Order Reactions,"
    ORNL-2958 (October 1960).
14. Chapman, T. G. HFIR Target Design Study, ORNL-TM-1084
    (September 1965).
15. Ferguson, D. E.; and Bigelow, J. E. "Production of ^{252}Cf
    and other Transplutonium Isotopes in the United States of
    America," Actinides Rev., 1, 213–221 (1969).
16. Crandall, J. L. "Tons of Curium and Pounds of Califor-
    nium," Proc. Conf. Constructive Uses of Atomic Energy,
    Washington, D.C., p. 193 (1968).
17. Seaborg, G. T.; Crandall, J. L.; Fields, P. R.; Ghiorso, A.;
    Keller, O. L.; and Penneman, R. A. "Recent Advances in
    the United States on the Transuranium Elements," in Proc.
    U.N. Intern. Conf. Peaceful Uses Atomic Energy, 4th Geneva
    2, 4.6-1 (1971).
18. Lotts, A. L.; Adams, R. E.; Bigelow, J. E.; King, R. T.;
    Long, E. L. Jr.; Manthos, E. J.; Van Cleve, J. E. Jr.
    Analysis of Failure of HFIR Target Elements Irradiated
    in SRL and HFIR – An Interim Status Report, ORNL-TM-2236,
    (February 1972).

RECEIVED February 24, 1981.

The Production of Transplutonium Elements in France

G. KOEHLY, J. BOURGES, C. MADIC, R. SONTAG, and C. KERTESZ

Commissariat à l'Energie Atomique, Centre d'Etudes Nucléaires—Section des Transuraniens, 92260 Fontenay-aux-Roses, France

Over a period of several years, the French Commissariat à l'Energie Atomique (C.E.A.) has developed a program for the production of transplutonium elements in order to satisfy its own requirements and also to supply the market for manufactured products such as ionizing sources (smoke detectors, lightning conductors) and gamma and neutron sources.

The isotopes are either produced by special irradiation of appropriate targets (plutonium 239/aluminum) in the case of americium 243 and curium 244, or, for americium 241, recovered from industrial wastes produced by reprocessing plants and plutonium oxide recycling. The annual production required to satisfy the various needs are respectively :

- ^{241}Am a few hundred g/year,
- ^{243}Am 15 g/year,
- ^{244}Cm 15 g/year.

All the chemical purification operations for the various isotopes are performed at the Fontenay-aux-Roses Nuclear Research Center by the Section des Transuraniens (S.T.U.) which has a group of hot cells adequate for these production operations.

To minimize corrosion of stainless steel equipment all the processes use nitric acid solutions. Hence polyaminoacetic acid complexing agents such as DTPA are required to accomplish the critical problem of the separation of trivalent actinides from trivalent lanthanides (1, 2).

The processes developed initially were based essentially on liquid-liquid extraction techniques, but the chemical problems encountered in the treatment of irradiated Pu/Al targets (e.g. considerable interface fouling in the extractors and formation of stable emulsions) and the intensification of safety requirements led to use of extraction chromatographic techniques.

Experimental

Sources of transplutonium elements. The main characteristics of the irradiated plutonium 239/aluminum targets containing
the isotopes americium 243 and curium 244, and the active solutions containing americium 241 are summarized in Tables I and
II.

Table I
Characteristics of Irradiated Pu/Al Targets.

FUEL ELEMENT
        total mass   5445 g
        mass of plutonium 239   400 g
        dimensions   1067 x 79.7 x 67.1 mm
        No. of plates   11

IRRADIATION, COOLING
        reactor   Celestin, (Marcoule)
        integrated flux   11.28 n.kb^{-1}
        cooling   3 years

COMPOSITION AFTER COOLING
        actinide mass :
        • ^{242}Pu 44 g
        • ^{243}Am 8.5 g
        • ^{244}Cm 7.5 g
        fission products :
        •   total mass   340 g
        •   rare earths   ≃240 g
        •   activity β,γ $3.7.10^{4}$ Ci.

The Pu/Al targets were initially the fuel elements for the
Célestin reactor. After irradiation and cooling, some targets
were   re-irradiated   for   nine   months   (average   flux
2.5 x 10^{14} n.cm^{-2}.s^{-1}) to improve the isotopic quality of the
americium 243 and plutonium 242 which they contained.

The active "Masurca" solution (Table II) is a special type
of waste resulting from the reprocessing of certain irradiated
fuels and from criticality analyses.

Cadmium was added to the solution for safety reasons, and
then its initial volume of about 400 m^3 was reduced to 4 m^3 by
distillation.

The second waste solution is produced by the reprocessing
of fabrication scrap from fabricating (U, Pu)O$_2$ fuels intended
for fast breeder reactors (3). The annual volume of this waste
solution amounts to some tens of m^3.

Table II
Characteristics of Waste Solutions

COMPONENT		MASURCA	PuO$_2$ reprocessing waste
HNO$_3$ (N)		1.1	4.9
U		12	0.005
Np		0.18	
Pu		0.07	0.005
^{241}Am		0.108	0.028
Fe	g/L	11.1	*
Cd		35.4	
Ni		1.01	*
Cr		1.5	*
^{144}Ce		0.28	
^{106}Ru	mCi/L	1.34	
^{137}Cs		9.1	

* Very Low Concentration present.

Hot Cells. For its transplutonium element production program, the S.T.U has a series of seven hot cells, the main characteristics of which are given in Table III. The Petrus cell (4, 5) (Figure 1) occupies a pivotal position among the various operations carried out including receipt of irradiated targets and active solutions transported in transfer casks ; storage of these materials ; storage and removal of liquid wastes from different cells and storage, packaging and removal of solid wastes. In addition to the "support" functions carried out for other hot cells, a number of process operations are also performed in Petrus, namely dissolution of irradiated targets and chemical treatment of irradiated targets including extraction of plutonium 242 and the first (americium 243, curium 244)/lanthanides separation cycle. Research related to other programs is also conducted in the Petrus cell.

The overall organization of the production of the isotopes americium 241, americium 243 and curium 244, showing the function of each hot cell as well as their interconnections, is shown schematically in Figure 2. Transfers can be made from the hot cells to the Petrus cell as follows :

. for liquid in double-jacket pipes (Pollux, Irene, Pétronille I),

. for analytical samples by pneumatic transfer (Pollux, Pétronille II),

. for solids and solutions rich in transplutonium elements by Padirac (6) casks protected by 10 or 15 cm of lead.

TABLE III
Main Characteristics of the Hot Cells of the S T U

CHARACTERISTICS	Name of Hot Cell						
	PETRUS	POLLUX	CANDIDE	PETRONILLE I	PETRONILLE II	IRENE	ANTINEA
Outer Dimensions							
Length –	15 m	8.1m	7.1m	3.90m	3.90m	3.45m	10.1m
Width –	5.1m	1.8m	1.7m	1.45m	1.45m	1.45m	2.6m
Height –	5.1m	2.5m	2.5m	2.00m	2.00m	2.37m	2.9m
Biological Shield							
.type	αβγη	αβγ	αβγ	αβγ	αβγ	αβγ	αη
	Spec. Concrete Pb	Pb	Pb	steel	steel	Pb	water
.thickness	1m	15cm	15cm	10cm	10cm	5cm	80cm
Solution Storage Capacities	5.3m^3	1m^3	70L	150L	150L	180L	0
Work Stations	7 front 4 rear	5	4	2	2	2	4
Telemanipulators							
Number	front rear / 14 3/1*	10	8	4	4	8	8
Model	G F	M7	MAll	M7	M7	MAll	MAll
Manufacturer	CRL *	CRL	La Calhene	CRL	CRL	La Calhene	La Calhene
inputs/outputs							
.Padirac (6)	yes	yes	yes	yes	yes	no	no
.IL 48 or 22**	yes	no	no	no	no	no	no

* 3 CRL, 1 Wallish-Miller
** Heavy transfer casks

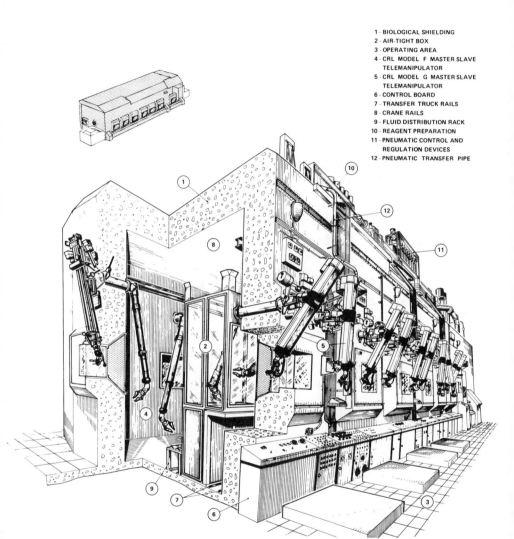

1 - BIOLOGICAL SHIELDING
2 - AIR-TIGHT BOX
3 - OPERATING AREA
4 - CRL MODEL F MASTER SLAVE
      TELEMANIPULATOR
5 - CRL MODEL G MASTER SLAVE
      TELEMANIPULATOR
6 - CONTROL BOARD
7 - TRANSFER TRUCK RAILS
8 - CRANE RAILS
9 - FLUID DISTRIBUTION RACK
10 - REAGENT PREPARATION
11 - PNEUMATIC CONTROL AND
      REGULATION DEVICES
12 - PNEUMATIC TRANSFER PIPE

Figure 1. Petrus—general view

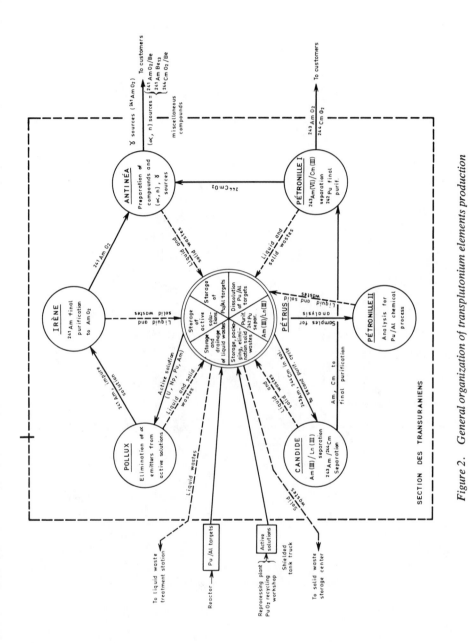

Figure 2.   General organization of transplutonium elements production

Previous experience in the production of plutonium 238 revealed the need for the double alpha containment of the cells where the alpha-emitter isotopes with high specific activity are handled. Thus all the hot cells are equipped with a double ventilation system which provides ventilation of alpha-cells and ventilation between alpha-cells and biological shields. Alpha detectors continuously monitor the exhaust circuits.

Operations designed to carry out production are distributed among the different hot cells in accordance with the biological shielding required and the storage capacity for solutions and liquid wastes. Thus for the treatment of irradiated Pu/Al targets, the "very high beta-gamma activity" cycle is performed in Petrus, which has large solution storage capacities. The "medium beta-gamma activity" is carried out in Candide, in which the beta-gamma shielding is less than that of Petrus (15 cm Pb). The final purification operations which practically involve alpha-emitters only are carried out in Petronille I with low beta-gamma shielding.

In the case of the treatment of waste solutions for the recovery of americium 241, the distribution of tasks was based essentially on storage volume requirements. The first purification cycles were performed in Pollux which has capacity for storing large volumes of solutions ; final purification of americium 241 was performed in Irene.

The correct management of the various gaseous, liquid and solid wastes determines the regularity of production. The gaseous wastes flow through exhaust systems equipped with absolute filters and soda lime and activated charcoal traps. They are discharged into the atmosphere after analysis of their krypton 85 and iodine content. The liquid wastes from the various cells are stored in 1 m^3 tanks (located in Petrus) and removed periodically by a shielded tank truck for transport to a liquid waste treatment station in another C.E.A. center.

The solid wastes are packaged in "La Calhène" type polyethylene bins (7). These bins are then placed in steel drums and embedded in a quick-setting resin. After the covers are crimped, the drums are removed by shielded transfer casks. To provide for direct access to final alpha-waste storage, it is planned to build a hot cell designed to treat solid wastes contaminated with alpha-emitters. The unit operations to be performed on these wastes will be storing, crushing, leaching, drying, and embedding in a cement/asphalt mixture.

Dissolver. The irradiated Pu/Al targets are dissolved in nitric acid in a dissolver placed in station 1 of the Petrus cell. This dissolver includes a reactor, bubble trap, condenser and soda scrubbing column for uncondensable gases. The reactor, built of Uranus 55 stainless steel, has an effective capacity of about 50 L.

Heating is provided by a heating fluid flowing through a double jacket, and the interior is lined with two coils designed to perform a cooling function (immersed coil) or a foam-breaker function (exposed coil). The dissolution of an irradiated Pu/Al target requires two operations and leads to the preparation of 88 L of solution.

Storage tanks. The total capacity of the storage tanks of each hot cell is given in Table III. The process tanks are double-jacketed for all the hot cells except Candide. This cell, designed initially for non-aqueous investigations, has small storage tanks placed directly on the alpha-cell work level.

Extraction units. The extraction chromatography columns are built of plexiglas, the solid stationary phase being immobilized in the column between two sintered glass discs. The caracteristics of the columns employed are given in Table IV.

TABLE IV

Characteristics of Extraction Chromatography Column

Stationary Phase Type [a]	Mass (kg)	Diameter (mm)	Effective Height (mm)	Void Volume (L)	Total Mass (kg)
$TOAHNO_3/SiO_2$	1.6	60	680	0.8	2.7
TBP/SiO_2	2.8	80	680	1.4	4.2
$POX.11/SiO_2$	9	150	680	4.5	29
$HD(DiBM)P/SiO_2$	0.7	60	300	0.35	1.3

(a)
$TOA.HNO_3$ = Trioctylamine nitrate
TBP = Tributylphosphate
POX.11 = Di-n-hexyloctoxymethylphosphine oxide
HD(DiBM)P = Bis-2,6-dimethyl-4-heptyl phosphoric acid

The solution is fed to the columns mainly by Prominent type (West Germany) proportioning pumps, except in the case of columns packed with the mixture $HD(DiBM)P/SiO_2$, for which the solution is fed by a pressurized tank. For the treatment of Masurca solution, uranium is extracted in macroconcentration by liquid-liquid extraction in mixer-settler batteries, in which the phases are in countercurrent flow. Two plexiglas batteries are employed ; one of them, a 10-stage battery used for uranium extraction, is designed for a total solution flow rate of 10 L/h, and the second 8-stage battery, employed for uranium stripping, allows a total flow rate of 5 L/h.

In-line detection. In the treatment of solutions containing americium the americium 241 content of solutions leaving the chromatography columns is continuously monitored. The gamma detector is a miniature Geiger-Muller tube placed in contact with the solution outlet pipe. The unit is shielded from ambient irradiation by lead.

Preparation of stationary phases. The stationary phases previously used (8, 9), Célite 545 and Gas Chrom Q (Applied Science Laboratory) were replaced for operating convenience and cost reasons by silica gel marketed by Merck (West Germany). This is a silanized 60 to 230 mesh stationary phase which also offers the advantage over previous stationary phases of being more dense, making it possible, with equivalent geometries, to achieve chromatographic columns with higher exchange capacity.

The silica gel was impregnated by the extractants TOA, TBP, POX.11 and HD(DiBM)P. Silica gel is placed in contact with a solution of extractant in hexane or acetone and the solvent is then evaporated under reduced pressure in a Buchi Rotavapor type rotary evaporator. The production capacity is 3 kg of stationary phase loaded with extractant per day. The mass extractant impregnation ratios in the final mixture are : TOA(25 %), TBP(27 %), POX.11(30 %), HD(DiBM)P(30 %).

Reagents. The reagents HNO_3 ; $Al(NO_3)_3 . 9H_2O$; $LiOH.H_2O$ (Prolabo), dodecane (Progil) are of technical grade quality while $K_2S_2O_8$; $AgNO_3$ (Prolabo), DTPA (K and K Laboratories), EDTA (Merck), TBP (Osi), TOA (Fluka) are pure materials of analytical grade quality. POX.11 is synthesized at our request by the IRCHA (91 VERT LE PETIT, France) ; HD(DiBM)P is synthesized in our laboratory by the method described in (10). The aluminum nitrate solutions deficient in NO_3^- ions, of the formula $Al(NO_3)_{3-x}(OH)_x$, are prepared by the destruction of NO_3^- ions at 100/120°C by formaldehyde (using 2 moles of formaldehyde per NO_3^- ion to be destroyed).

Results and Discussion.

Treatment of irradiated targets. The chemical operations relative to the production of transplutonium elements (americium 243, curium 244) are all performed using a nitric acid medium. The highly corrosive nature of the solutions concentrated with Cl^- ions, which were used in the USA for the development of the Tramex process (11), and the instability of SCN^- ions to radiation (12), led us to select nitric acid solution to perform the chemical separations. Once the medium was selected, it was necessary to find an adequate additive which, in combination with a suitable extractant, would allow solution of the main problem namely separation of the trivalent actinides from trivalent lanthanides.

The family of polyaminoacetic complexing agents, especially diethylene triaminopentaacetic acid (DTPA), helps to obtain this objective. Table V summarizes some separation factors obtained for systems all using DTPA in combination with the extractants HDEHP ($\underline{13}$), TBP ($\underline{2}$) and TLAHNO$_3$ ($\underline{2}$).

While the Talspeak system allows the best separation of the group, systems using TBP or TLAHNO$_3$ allow good separation of americium with light rare earths which, as is well known, make up the bulk of the mass of lanthanides present in the irradiated targets ($\underline{14}$).

An initial experiment involving the treatment of small irradiated Pu/Al targets for the production of americium 243 and curium 244 was carried out in France in 1968 ($\underline{2}$). The chemical process was based essentially on the use of a system comparable to the Talspeak system. After plutonium extraction by a 0.08 M trilaurylammonium nitrate solution in dodecane containing 3 vol % 2-octanol, the actinides (americium, curium) were co-extracted with a fraction of the lanthanides by a 0.25 M HDEHP - dodecane solvent from an aqueous solution previously neutralized by Al(NO$_3$)$_{3-x}$(OH)$_x$ and adjusted to 0.04 M DTPA. The actinides were selectively stripped by placing the organic phase in contact with an aqueous solution of the composition 3 M LiNO$_3$ - 0.05 M DTPA. While this experiment achieved the recovery of 150 mg of americium 243 and 15 mg of curium 244 with good yields, the process presented a drawback due to the slow extraction of Al(III) which saturates the HDEHP. This process was therefore abandoned.

The treatment scheme for the first irradiated targets ($\underline{8}$) was based on the TLAHNO$_3$/DTPA system implemented by liquid-liquid extraction. After dissolution of the Pu/Al targets by nitric acid, the solution was adjusted to low acidity by addition of Al(NO$_3$)$_{3-x}$.(OH)$_x$ and then countercurrently contacted with an organic solution of the composition 0.64 M TLA.HNO$_3$ in dodecane containing 3 vol % 2-octanol. The co-extracted elements are then separated by selective stripping as follows :

a.   Stripping of americium 243 and curium 244 by placing the loaded organic solution in contact with an aqueous solution of the composition 6 M LiNO$_3$ - 0.1 $\underline{M}$ DTPA at pH = 2.

b.   Stripping of lanthanides by a 2 M HNO$_3$ solution.

c.   Stripping of plutonium by an aqueous solution of the composition 0.5 M H$_2$SO$_4$; 0.2 N HNO$_3$; 0.05 M Fe^{2+}.

TABLE V

Separation Factors for Some Chemical Systems Using DTPA[a]

System

Ln(III)	HDEHP (Talspeak)		TBP		TLAHNO$_3$	
	α_i	$\alpha_{(DTPA)}$	α_i	$\alpha_{(DTPA)}$	α_i	$\alpha_{(DTPA)}$
La	(3.9)	380	0.80	1800	6.0	2300
Ce	5.4	140	0.86	52	3.5	660
Pr		(75)	1.06	12.5	2.3	73
Pm	(10)	(75)	1.1	5.8	1.2	9
Eu	48	91	1.3	2.0	0.83	1.13
Tb			1.5	1.4	0.43	0.90
Er			1.02	1.4	0.20	0.43
Tm		(8000)	0.93	1.4	0.19	0.33
Yb			0.74	1.4	0.17	

$(\alpha) = D(LnIII)/D(AmIII)$.

(a)
Operating conditions :
- Talspeak :
  - Organic phase   0.2 M HDEHP in Di-isopropylbenzene
  - Aqueous phase   1 M lactic acid, pH = 3(α_i) + 0.05 M DTPA (α_{DTPA})
- TBP :
  - Organic phase   TBP (40 % vol) in dodecane
  - Aqueous phase   4 M LiNO$_3$(α_i) + DTPA = Al^{3+} = 0.25 M (α_{DTPA})
- TLAHNO$_3$:
  - Organic phase   TLAHNO$_3$(40 % vol) in dodecane (50 %) - chlorobenzene (50 %)
  - Aqueous phase   6 M LiNO$_3$(α_i) + DTPA = Al^{3+} = 0.25 M (α_{DTPA})$_3$.

Although the process proved satisfactory from the chemical standpoint, practical problems emerged in that the hydraulic operation of the mixer-settler batteries was extremely poor. In effect, as soon as the aqueous solutions from the dissolution of irradiated targets were placed in contact with the organic extraction phases, a stable emulsion was formed, produced by the appearance of extensive precipitates at the aqueous solution/organic solution interface. As no chemical remedy was found to solve this problem, we attempted to adapt this type of process to extraction chromatographic techniques.

The second problem raised by the production of americium 243 and curium 244 resides in their mutual separation. The first experiment developed in France ($\underline{1}$) was based on a chromatographic Am/Cm separation on Dowex 1 x 8 anion exchange resin (200/400 mesh) in NO_3^- form. After the fixation of Am(III) on the resin from an ethanolic (80 vol %) solution of composition 1.33 M NH_4NO_3; 0.1 M HNO_3, the curium was eluted with an ethanolic solution of 1.33 M NH_4NO_3 - 0.025 M DTPA, while the Am(III) was eluted by 1 M HNO_3 or with an ethanolic (80 vol %) 1.33 M NH_4NO_3 - 0.025 M DTPA solution. This technique, which allows good Am/Cm separation, is difficult to implement in a hot cell because of safety reasons related to the presence of ethanol. The second method developed ($\underline{2}$) was based on the selective stripping of Cm(III) from an 0.64 M TLA.HNO_3 in 3 vol % octanol-2 - dodecane by an aqueous solution of the composition 4.25 M $LiNO_3$ - 0.1 M DTPA - 0.1 M Al whose acidity was adjusted to obtain $YH_3^{2-} = YH_2^{3-} = 0.05$ M (with YH_5 = DTPA).

In these conditions, α = D Am(III)/D Cm(III) is 3.25. The Am/Cm separation required the use of two batteries of mixer-settlers of twenty stages each, one assigned to Cm(III) stripping and the second to scrubbing of the aqueous phase containing curium. At the outlet of the second battery, the organic scrubbing solution (containing small amounts of americium) was sent to the first battery, where it diluted the organic phase loaded with (Am, Cm), making up the bulk of the organic solution for this battery. This system proved effective but contained two drawbacks : the sensitivity of Am(III) distribution coefficients to the ratio (DTPA/Al^{3+}) of the aqueous solution, and the difficulty of keeping the proportioning pump deliveries constant to within $\pm$ 3 %, which resulted in alteration of the extraction factors. Hence this process was not adopted for the treatment of irradiated targets.

Americium can exist in oxidation states higher than III, this chemistry is known to have been used for Am/Cm separation as follows :

a.   Oxidation to Am(V) by $S_2O_8^=$ ions in concentrated K_2CO_3 medium, leading to the precipitation of $K_{2x-1}AmO_2(CO_3^=)_x$, while Cm(III) remains in solution ($\underline{15}$).

b. Oxidation of the americium to Am(VI) by $S_2O_8^=$ ions in slightly acidic solution (0.1 M HNO_3), followed by precipitation
of CmF_3 by the addition of F^- ions (16).

These two techniques are routinely used for the Am/Cm separations that we perform. However, they both have disadvantages.
In the case of the precipitation of the double carbonate, the
relatively high solubility of americium in solution (50 to
100 mg/L) requires reprocessing of the supernatant solution from
the precipitation step which must be diluted considerably to
avoid the precipitation of KNO_3 after acidification by HNO_3. For
method (b), failure to oxidize the americium results in a loss to
the CmF_3 precipitate.

Furthermore, this method is not reasonably applicable to
Am/Cm mixtures in which curium is present in macroconcentration.
Recently (9), we adapted the separation method developed by
Mason, Bollmeier and Peppard (10) to handle macroconcentrations
of Am. This method consists of a selective extraction of americium, after its oxidation to Am(VI), by an extractant with
outstanding selectivity properties, HD(DiBM)P. It can be implemented either by liquid-liquid extraction, requiring the use of
centrifugal extractors (17), or by extraction chromatography, a
simpler and less costly technique.

The problem of recovering the plutonium contained in the
Pu/Al target dissolution solutions is trivial in comparison with
the difficulties discussed above. The strong affinity exhibited
by tertiary amine nitrates (TLA or TOA) for Pu(IV) was exploited
to develop the following processes :

a. Final purification of plutonium as part of the reprocessing
    of irradiated fuels (18).

b. Treatment of irradiated neptunium 237 targets for the pro-
    duction of plutonium 238 (19).

Plutonium extraction requires a prior adjustment to the
(IV) oxidation state which can be accomplished by H_2O_2. Pu(IV)
can be stripped from the columns packed with the mixture
$TOAHNO_3/SiO_2$ by a solution of the composition 0.75 M H_2SO_4 -
0.2 M HNO_3.

Treatment of waste solution. The objective of the treatment
of wastes of the type described in Table II is twofold : first,
the elimination of alpha-emitters from the waste, and secondly
the recovery of americium 241 which can be utilized directly.
Since all the waste solutions contain nitric acid, the only
parameters which can conveniently be defined are :

a. Choice of the extraction system, and

b. Aqueous phase salting strength and complexant concen-
    tration.

The Masurca waste exhibits the special property of high UO_2^{2+} concentration (12 g/L), so that the first operation to be performed is TBP extraction of U(VI). This operation does not require prior feed adjustment, because the acidity and nitrate ion concentration are sufficiently high. The U(VI) concentration of the PuO_2 recycle waste, even after volume reduction by a factor of 50, does not require a specific uranium extraction cycle.

The critical point of the process lies in the definition of the americium 241 extraction conditions. We decided to use POX.11 which has been recommended for some years for this type of separation (20). The chief advantage offered by phosphine oxides over the more standard extractant TBP is that it allows the use of solutions with lower salt concentrations. In addition, slightly stronger nitric acid concentration can be tolerated (about 0.1 N), thus eliminating the risk of hydrolysis of certain metallic cations such as Fe(III). Also POX.11 offers the advantage over TOPO of being liquid and miscible in all propor-tions with standard solvents. Under the conditions of extraction of Am(III) by POX.11, Pu(IV) and neptunium are co-extracted. Extraction of the neptunium, which is initially present in the Np(V) form, is certainly due to the disproportionation to Np(IV) and Np(VI), which have a strong affinity for the extractant. The POX.11 nevertheless has the disadvantage of extracting Fe(III) with rather slow kinetics, which, in the case of the Masurca waste, gives rise to a sharp drop in $D_{Am(III)}$ as shown in Figure 3. The addition of EDTA to the aqueous solution in equal concentration to Fe(III) helps to avoid its extraction and to restore all the extractive properties of the POX.11 for Am(III) as shown in Figure 4. This is due to the selective formation of the complex FeY^- (with YH_4 = EDTA) whose formation constant $\log (FeY^-) = 25.1$ (21) is much higher than that of the complex $\log (AmY^-) = 18.0$ (22).

However, addition of EDTA to the waste requires the prior extraction of Pu(IV) and Np(IV), which form strong complexes with EDTA. This separation may be carried out simply, after adjusting the acidity and nitrate ion concentration by passing the solution through a column packed with an anionic exchange resin such as IRA-400.

Stripping of the Am(III) from an organic phase or a statio-nary phase loaded with POX.11 can be carried out by contact with an aqueous 6 M HNO_3 solution. Final purification of americium 241, in the case of Masurca waste, can be carried out by selec-tive extraction with TBP from a strongly salted (e.g., 7 M NO_3^-) and slightly acidic (e.g., 0.05 M HNO_3) solution, followed either by precipitation of $K_{2x-1} AmO_2(CO_3)_x$ or selective extraction of Am(VI) by HD(DiBM)P. Purification of americium 241 resulting from PuO_2 waste reprocessing is simpler. After strip-ping of the POX.11, the americium can be precipitated directly in oxalate form, which must remove traces of Fe(III) accompa-nying americium through this stage of the process.

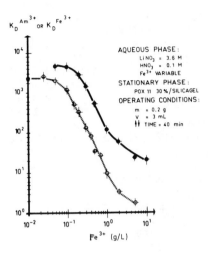

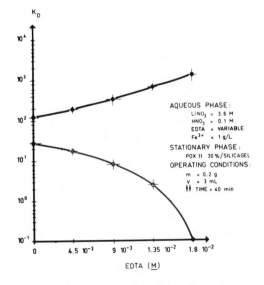

Figure 3. Extraction of Fe^{3+} and Am^{3+} by POX.11 30%/silicagel: variation of the distribution coefficients of (●) Am^{3+} and (○) Fe^{3+} with Fe^{3+} concentration

Figure 4. Extraction of Am^{3+} and Fe^{3+} by POX.11 30%/silicagel in a complexing medium: variation of the distribution coefficients of (●) Am^{3+} and (○) Fe^{3+} with EDTA concentration

Extraction techniques.

The technique employed almost exclusively in the process
developed is extraction chromatography (LLC), which is an ideal
technique, in view of the column exchange capacities, for the
extraction of moderate amounts of material present in a limited
volume, or for the recovery of elements present in low concentration in large volumes. The simplicity of the equipment and
the low sensitivity of extraction performance to fluctuations in
 solution throughputs make it a simpler technique for use than
liquid-liquid extraction. Moreover, LLC lends itself well to
discontinuous operation. In the case of difficult separations,
such as Am/Cm separation, LLC offers the advantage of a large
number of stages in a simple, compact unit. Since the extraction
of the U(VI) present in the Masurca waste in macroconcentration
is not economically feasible by LLC, the conventional liquid-
liquid technique was adopted.
Another conventional technique employed is that of ion
exchange chromatography on a column used in the Masurca process
for the co-extraction of neptunium and plutonium.

Process flowsheets. The flowsheets of the chemical processes used for the treatment of Pu/Al targets and the Masurca waste
were described recently (9, 23), so that we shall only discuss
the major principles here. However, we shall dwell in greater
detail on the process flowsheet adopted for treatment of the
waste produced by the recycling of PuO_2 scrap, which is expected
to go on stream in August 1980.

Treatment of Pu/Al targets. The basic steps of this treatment process are shown in Figure 5. The critical operations of
the process, Am(III)/Ln(III) and Am(III)/Cm(III) separations
both performed on TBP column in line with the same operating
scheme require two cycles. The adjustment of An(III)-rich
eluates is carried out simply by adding aluminum nitrate, which
displaces the An(III) ions from their complexes with DTPA.
Volume reductions are necessary before implementing the second
cycles, and are carried out by the fixation of An(III) on a small
TBP column, followed by elution with 8 M $LiNO_3$ - 0.1 M DTPA
solution of pH 5. The upstream use of filter columns packed with
silica gel significantly improved the fixation fronts. In addition, the column sizes and hence exchange capacities were
increased.
The technique for final americium purification was significantly changed in relation to the one described previously (9).
The "Am(VI)/F⁻" technique was abandoned for the reasons given
above in favor of the following method.

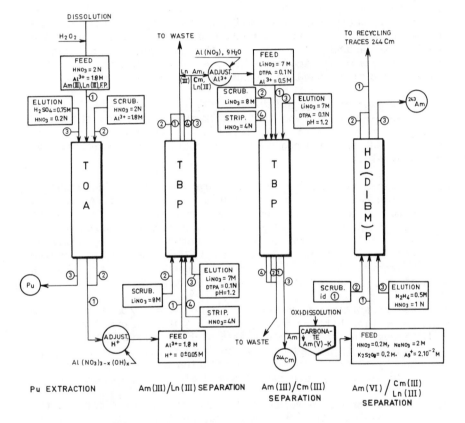

Figure 5. Chemical process for Pu/Al targets treatment

After changing the medium (allowing the elimination of
DTPA) on the TBP column, the americium containing small amounts
of Cm(III) and Ln(III) (Eu, Gd, Ce) is precipitated as an hydroxide, and redissolved in the minimum of nitric acid. This solution is then placed in 3.5 M K_2CO_3 medium and K_{2x-1} $AmO_2(CO_3)_x$ is
precipitated in the standard manner. This step already achieves
good separation of the impurities, although it is inadequate for
the americium to meet the required purity specifications. The
precipitate is redissolved by an oxidizing solution (see
Figure 5), leading to the production of Am(VI) which can be
injected directly into the HD(DiBM)P column. The use of this
technique served to cut the number of Am/Cm separation cycles
carried out on the TBP column, and thus to reduce significantly
the total operating time. The finished products, AmO_2 and CmO_2,
are obtained in a standard manner by the "oxalate" method.

Treatment of Waste Solutions Containing Americium. The process flowsheets developed to treat Masurca waste and the solution from recycling of PuO_2 scrap are presented in Figure 6.
Treatment of the Masurca waste is complex and requires the
following steps.

a. Liquid-liquid extraction of uranium by a TBP solution
    followed by stripping of the uranium with water. A low flow
    rate of an aqueous reducing scrubbing solution allows
    stripping of small amounts of Pu which are slightly co-
    extracted. The flow rate ratio between the feed solution
    and the stripping solution is 4.28, producing a uranium
    concentration of 51 g/L in the aqueous stripping solution
    at the outlet of the mixer-settler battery.

b. Co-extraction of Np and Pu : The waste from the uranium
    extraction battery is adjusted to 5 M $LiNO_3$ and then passed
    through an IRA-400 column at a high flow rate (30 L/h). The
    americium is not sorbed. Neptunium and plutonium are strip-
    ped by a dilute nitric acid solution and precipitated as a
    hydroxide, and calcined to yield mixture of oxides.

c. Extraction of [241]Am(III) on a POX.11 column : this is the
    main step of the treatment. Before introduction into the
    POX.11 column, the solution is adjusted to 0.1 to 0.15 M H^+
    and 0.16 M EDTA. This EDTA concentration corresponds to
    that of the Fe(III) present. This adjustment is the criti-
    cal step of the process, as a deficiency of EDTA in relation
    to Fe(III) results in co-extraction of Fe and Am in the
    column, thereby reducing capacity of the solvent for
    americium. Excess EDTA results in the presence of insoluble

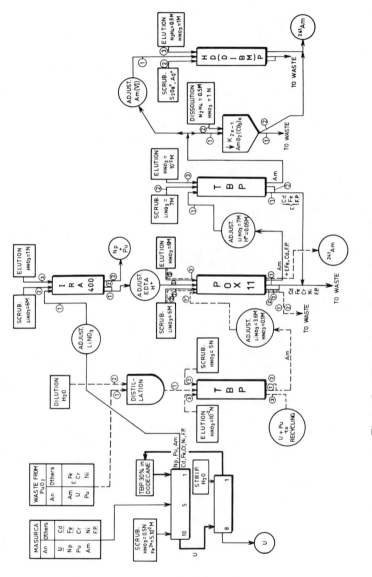

Figure 6. Chemical processes for industrial waste treatment

EDTA which cloggs filters and, sometimes, even the columns.
The feed pump is actuated by the signal of the gamma detector placed near the waste outlet pipe. This allows continuous feed of the active solution without the risk of
polluting the waste with americium 241. On the POX.11
column packed with 9 kg of stationary phase, one cycle
served to extract 15.2 g of americium 241 present initially
in 685 L of solution. The americium 241 concentration fact-
or was 34, and the americium 241 waste decontamination factor was 250. The americium eluate contained only very small
amounts of Fe and Cd. Decontamination factor of Am from Fe
and Cd were 130 and 140, respectively.

d. Am(III) extraction on TBP column : The extraction of
    Am(III) requires prior adjustment of the solution to 0.05 M
    HNO_3 and a high concentration of $LiNO_3$. Under these condi-
    tions, the Fe(III) and Cd(II) present in concentrations
    comparable to that of americium, are not extracted. Never-
    theless, when the acidity of the feed solution is lower
    than 0.05 M, a "pseudo extraction" of Fe(III) is noted. A
    rust-colored band ($Fe_2O_3.x\ H_2O$) appears on the column, and
    migrates towards the outlet at a lower speed than the
    liquid phase. This phenomenon can be explained by the
    extraction of nitric acid by TBP causing hydrolysis of
    Fe(III) followed by redissolution of the precipitate by the
    HNO_3 in the feed solution ; these steps are repeated
    gradually along the column.

e. Final purification of americium 241 : Two methods are avai-
    lable, precipitation of $K_{2x-1}\ AmO_2(CO_3)_x$ or extraction of
    Am(VI) on a HD(DiBM)P column. In both cases, the americium
    from the TBP column is first precipitated as $Am(OH)_3$.

f. Production of AmO_2 : The americium from one of the final
    purification methods undergoes reduction to Am(III). It is
    then precipitated as $Am(OH)_3$ which is then dissolved in
    HNO_3. Subsequently, americium is precipitated as
    $Am_2(C_2O_4)_3.10\ H_2O$ which is calcined at 800°C for three
    hours to give AmO_2.

The conceptual flowsheet for treatment of the waste from PuO_2
recycle operation is less complex. The operations to be
performed are the following.

a. Volume reduction : The waste is concentrated by a factor of
    50 by distillation in glass equipment. The concentrate thus
    obtained is about 11.3 M HNO_3.

b. Co-extraction of Pu(IV), Pu(VI) and U(VI) on TBP column :
    The concentrate, diluted with water to adjust the acidity
    to 5 M HNO_3, is injected into the TBP column. The slightly
    acidic eluate containing Pu(IV), Pu(VI) and U(VI) is re-
    acidified to 1 M to prevent hydrolysis of the Pu(IV).

c. Extraction of ^{241}Am(III) on POX.11 column : The waste from fixation and scrubbing of the TBP column is adjusted to a 3.6 M LiNO$_3$ - 0.1 M HNO$_3$ solution. The small amount of Fe(III) present does not require complex adjustment with EDTA. But the Am product also contains a small amount of the iron initially present in the waste.

d. Am/Fe separation : The precipitation of Am(III) oxalate allows good separation of the Fe(III). If necessary, a second precipitation is carried out.

Developments in processes and programs.

The major drawback of the processes described here obviously lies in the generation of liquid wastes heavily loaded with nitrates salts, thus complicating their subsequent management. For some wastes, it is impossible to avoid this high salt concentration. In the case of the treatment of Pu/Al targets, these wastes contain the fission products ; their salt concentration (Al^{3+}) is imposed by the nature of the target. For the Masurca waste, the initial solution is already heavily loaded with nitrate salts.

A two-fold reduction in the salt concentration can be achieved in the treatment of Pu/Al targets by substituting a phosphine oxide such as TOPO for the TBP as shown by a recent study (24). In the case of treatment of the waste from PuO$_2$ recycling, it is possible to avoid the presence of salt completely, provided that a polydendate extractant exhibiting a strong affinity for Am(III) in acidic medium, such as DHDECMP, is used (25), (26). While this system appears to be attractive for extraction of Am(III) from 4 to 6 N HNO$_3$ media and elution of Am(III) with water, it has so far found limited use.

Conclusions.

The development of the program for the production of transplutonium elements, americium 241, americium 243 and curium 244 in France required a major effort from the technological and chemical standpoints. Pre-existing hot cells were reconditioned and others were specially built for these production operations. From the chemical standpoint, the development of extractive chromatography on the preparative scale has allowed the definition of simple processes whose performance characteristics in our operating conditions have proved to be better than those obtained by liquid-liquid extraction. This type of process, initially developed for the treatment of Pu/Al targets, is ideal for the treatment of industrial wastes for their decontamination and for the production of americium 241.

Literature Cited

1.   Berger,R.;   Koehly,G.;   Musikas,C.;   Pottier,R.;
      Sontag,R. Nuclear Appl. Tech., 1970, 8, 371.
2.   Koehly,G.; Berger,R. "Symposium sur les Elements
      Transuraniens" : Liège 21-22 April 1969, 91.
3.   Arnal,T.; Cousinou,G.; Ganivet,M. Report CEA-R-4946.
4.   Sontag,R.; Koehly,G.; Report CEA-R-4470.
5.   Berger,R.; Faudot,G.; Sontag,R. Report CEA-R-4471.
6.   Vertut,J.; Lefort,G.; Brissac,M.; Cazalis,J.P. Proc.
      17th Conf. Remote Syst. Tech. Am. Nucl. Soc., 1969,
      165.
7.   Lefort,G.; Vertut,J.; Cazalis,J.P. Proc. 11th Conf.
      Hot. Lab. Equip. Am. Chem. Soc., 1963, 353.
8.   Bourges,J.; Koehly,G. B.I.S.T., 1976, 218, 67.
9.   Bourges,J.; Madic,C.; Koehly,G. A.C.S. Symposium
      Series, 1980, 117, 33.
10.   Mason,G.W.; Bollmeier,A.F.; Peppard,D.F.; J. Inorg.
      Nucl. Chem. 1970, 32, 1011.
11.   Leuze,R.E.; Baybarz,R.D.; Weaver,B. Nucl. Sci. Eng.
      1963, 17, 252.
12.   Moore,F.L.; Anal. Chem., 1964, 36, 2158.
13.   Weaver,B.; Kappelman,F.A.; J. Inorg. Nucl. Chem.,
      1968, 30, 253.
14.   Orth,D.A.; Mc Kibben,J.A.; Prout,W.E.; Scotten,W.C.
      Proc. Int. Solvent Extr. Conf. The Hague. Society of
      Chemical Industry, London, 1971.
15.   Burney,G.A. Nucl. Appl., 1968, 4, 217.
16.   Proctor,S.G. J. Less Com. Metals, 1976, 44, 195.
17.   Musikas,C.; Germain,M.; Bathellier,A. A.C.S. Sympo-
      sium Series, 1980, 117, 157.
18.   de Trentinian,M.; Chesné,A.; Report C.E.A. 1426, 1960.
19.   Berger,R.; Koehly,G.; Espié,J.Y. Proc. Int. Solvent
      Extr. Conf., Society of Chemical Industry, London,
      1971, 792.
20.   Guillaume,B. Personal communication.
21.   Gustafson,R.L.; Martell,A.E. J. Phys. Chem., 1963, 67,
      576.
22.   Moskvin,A.I.; Khalturin,G..V.; Gel'man,A.D. Radiokhi-
      miya, 1959, 1, 141.
23.   Madic,C.; Kertesz,C.; Sontag,R.; Koehly,G. "Symposium
      on Separation Science and Technology for Energy Appli-
      cations", Oct.-Nov. 1979, Gatlinburg, Tennessee.
24.   Kosyakov,N.V.; Yerin,E.A.; Vitutzev,V.M.; J. of
      Radioanal. Chem., 1980, 56, (1-2), 83.
25.   Schulz,W.W.; "The Chemistry of Americium" ERDA
      CRITICAL REVIEW SERIES, TID 26.971, 1976.
26.   Schulz,W.W.; Navratil,J.D.; "Actinide Separations",
      A.C.S. Symposium Series, 117, 1980.

RECEIVED March 13, 1981.

Transplutonium Elements, By-Products of the Nuclear Fuel Cycle

GÜNTER KOCH

Institut für Heisse Chemie, Kernforschungszentrum Karlsruhe, D-7500 Karlsruhe, Germany

WOLFGANG STOLL

ALKEM GmbH, D-6450 Hanau, Germany

A research and development program on the recovery and purification of potentially useful by-product actinides from the nuclear fuel cycle was carried out some years ago in the Federal Republic of Germany as part of the "Actinides Project" (PACT). In the course of this program, procedures for the recovery of neptunium, americium and curium isotopes from power reactor fuels, as well as procedures for the processing of irradiated targets of neptunium and americium to produce heat-source isotopes, have been developed. The history of the PACT Program has been reviewed previously (1). Most of the PACT activities were terminated towards the end of 1973, when it became evident that no major commercial market for the products in question was likely to develop.

Later, development work in this field was done primarily with the goal of removing transplutonium isotopes (specifically ^{241}Am) and neptunium from certain product and medium-active waste streams in order to meet product specifications, or to facilitate the handling of those streams. In this regard, neptunium tends to pose the more obvious obstacles in nuclear fuel reprocessing flow schemes, and procedures to improve decontamination from this element have therefore been quite intensely studied (2, 3). However, neptunium not being a transplutonium element, and therefore not fitting into the scope of this Symposium, this subject will not be further discussed in this paper. Problems with ^{241}Am in the fuel cycle originate mainly from its build-up by decay of ^{241}Pu during storage of plutonium. Because of its quite intense gamma emission, ^{241}Am can become a nuisance in plutonium fuels fabrication by direct or "hands-on" operations, and some decontamination from this isotope, before further processing of stored plutonium, may become necessary in order to reduce personnel exposure.

Recovery of Am and Cm from High-Level Waste

Americium and curium isotopes formed during irradiation of nuclear reactor fuels are diverted into the high-level waste (HLW) stream during fuel reprocessing. The HLW is thus the biggest

0097-6156/81/0161-0041$05.00/0

potential source for these elements, and activities to develop
a process for the recovery of Am and Cm from HLW were started
in 1967. Major guidelines were that the process to be developed
must not essentially increase the volume of waste to be processed
further, must not use strongly corrosive reagents, and must be
compatible with the final waste solidification procedure. The development of the recovery flowsheet, which was based on the extraction of a lanthanides - actinides fraction by di(2-ethylhexyl)
phosphoric acid (HDEHP) and on a "reverse-TALSPEAK" separation of
Am and Cm from the rare earths, has recently been reviewed(1), so
that a short description may be sufficient at this point.

The final version of the flowsheet (4,5,6) is given in Fig. 1.
The high-level waste (designated "1WW" in Fig. 1) is denitrated
with formic acid, with the goal (a) to reduce the acidity of the
HLW down to a value suitable for HDEHP extraction, and (b) to
remove "trouble-making" fission and corrosion products by precipitation, thus eliminating the need to add organic complexants to
the extraction feed. Conditions were chosen such that Am, Cm and
rare earths (R.E.) remained in solution while most of the Zr, Nb,
Mo, noble metals, and Fe were precipitated. The solid sludge could
be filtered off and the filtrate fed to the solvent extraction
cycle, using 0.3 $\underline{M}$ HDEHP (extractant) + 0.2 $\underline{M}$ TBP (modifier) in a
n-alkane diluent as the solvent. Am, Cm and R.E.'s were extracted
in the WA contactor from the non-complexed feed, Am and Cm were
partitioned from the R.E.'s in the WB contactor into 0.05 $\underline{M}$
diethylene triamine pentaacetate (DTPA)/1$\underline{M}$ lactic acid, R.E.'s
were re-extracted from the solvent in the WC contactor by 5 $\underline{M}$
nitric acid, and the Am/Cm product solution from WB was further
purified from R.E.'s by an additional organic solvent scrub stream
in the WD contactor. For the final purification and concentration
of the Am + Cm product a cation exchange process was developed.
Separation of the Am from Cm, if necessary, might be performed by
the Hanford cation exchange process (7,8), by high-pressure cation
exchange (9-11), or by potassium americium (V) carbonate precipitation (12); for reviews of these procedures see references 13 and
14.

In laboratory tests using simulated HLW solution spiked with
fission product tracers, Am and Cm, the denitration step proved
to be a sensitive process, but Am/Cm recoveries of ca. 90% in the
aqueous supernate could be realized under optimized conditions.
Decontamination factors (DF) > 1000 for Zr, Nb, Mo, and ∿ 100 for
Ru and Fe were obtained in the precipitation step. The solvent extraction cycle gave > 98% recovery of Am/Cm and DF > 10^3 for rare
earths, Sr and Cs. Appreciable decontamination was also obtained
for Zr/Nb (DF = 20), Ru (50), U (650), Pu (250), Np (800) and Fe
(420). The ion exchange cycle served mainly for Am-Cm concentration and for removal of DTPA and lactic acid; based on tests with
europium as a stand-in for trivalent actinides, concentration factors of about 50 could be expected under optimized conditions.

Planning of a pilot plant for the recovery of Am and Cm was
started . The name ISAAC (from the German "$\underline{I}$solierungs-$\underline{A}$nlage für

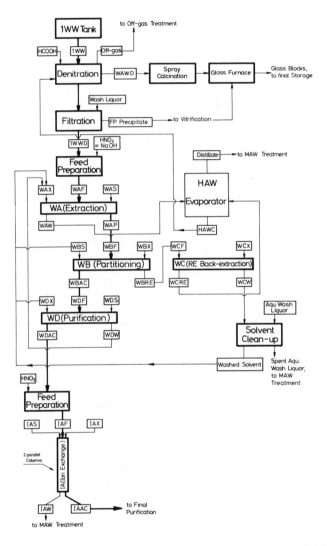

Figure 1.   Flowsheet for Am–Cm recovery from high-level waste (1WW) solutions (1). For a list of compositions and flow rates of the process streams see References 4 and 5.

Americium und   Curium) was given to this pilot project (4). These
plans were abandoned when the PACT project was terminated in 1973.
Thus, demonstration of the separation process with actual HLW is
lacking, and no judgement can be made on its performance.

Recovery of [241]Am from stored Plutonium

This work was started with the aim of recovering some of the
[241]Am which is formed during storage of plutonium, without impo-
sing any trouble to the fuel element fabrication process itself(1).
A possible source of material is the plutonium-fuel fabrication
scrap which at the ALKEM plant is collected for recovery of plutonium values by anion exchange. The effluent from the anion exchange column contains the americium, together with uranium,
corrosion products, residues from chemical reagents, and non-re-
covered plutonium. For the recovery of Am (and Pu) from the concentrated effluents, a process based on oxalate precipitation and
solvent extraction with tricapryl methyl ammonium nitrate, TCMAN
(nitrate form of Aliquat-336, a product of General Mills Inc.) was
operated for some time in a small-scale facility equipped with
pulsed glass columns (15,16). The concentrated effluents were adjusted to 6 to 7 $\underline{M}$ HNO_3, and the U and Pu were extracted in the
first column by 0.5 $\underline{M}$ TCMAN dissolved in Sovesso-100, a high-
boiling aromatic diluent produced by Exxon Co. U and Pu were back-
extracted in a second column into an acetic acid – hydroxylamine
sulfate solution. The effluent from the first column was saturated with oxalic acid and neutralized with ammonia to pH = 2.5.
A nearly white precipitate of Am and Ca oxalates was obtained
while most of the metallic contaminants (e.g., Fe, Cr, Al) remained in solution as stable oxalato complexes. The precipitate
was filtered off, dissoved in boiling concentrated nitric acid to
destroy the oxalate, neutralized with ammonia to pH = 2.5 to 3,
and the Am was extracted from the strongly salted aqueous ammonium
nitrate solution by 0.5 $\underline{M}$ TCMAN/Solvesso. The loaded organic solvent was scrubbed with concentrated ammonium nitrate solution,
and the americium was back-extracted with dilute nitric acid, precipitated as the oxalate, and converted into AmO_2 by calcination
at 800°C. Multi-gram amounts of [241]Am have been prepared with
this procedure, with Am purities > 99%.

Modification into Technical-scale Operation

Utilization of plutonium in early research and commercial
orders to fabricate thermal recycle and fast breeder fuels did
not coincide in timing with Pu availability from different sources.
The plutonium comes mainly from high-exposure light-water reactor
fuel reprocessing; extended storage of this Pu as a nitrate solution leads to [241]Am contents up to 3%. For hands-on operation with
this material it is necessary to reduce the Am content to about
0.5%. It was also necessary to minimize the liquid waste streams
from the plant. In designing a technical-scale process, it was

essential to both utilize an existing precipitator and avoid
flammable liquids for the main product stream.

The flow scheme of the process (17) is represented in Fig. 2.
The required throughput rate of 5 kg Pu/day is obtained in a
batch-type operation , where a 5 to 10% substoichometric oxalate
precipitation is performed by adding solid oxalic acid to a 3 $\underline{M}$
HNO_3 - 100 g/L $Pu(NO_3)_4$ solution at 80°C in about 2 hours. Up to
95% of the Pu is precipitated as uniform crystals of 20 μm average size and filtered. After washing and calcination, the average
analysis of this product shows less than 1000 ppm total metallic
impurities. When evaporating the filtrate to about 5% of its
original volume, nitric acid is recovered, and most of the oxalic
acid is destroyed. This results from sump temperatures of up to
123°C and the presence of Pu(VI).

The concentrated filtrate is adjusted to 7 $\underline{M}$ HNO_3 and passed
over Permutit SK anion exchange resin to fix the remaining Pu.
Plutonium is eluted with 0.6 $\underline{M}$ HNO_3, evaporated, and added to the
main Pu stream. Americium passing the resin bed together with
metallic impurities is evaporated to a solution containing 20 g/L
Am(III) nitrate and all the corrosion products and impurities from
storage and processing. This solution is saturated with solid
oxalic acid ($\sim$ 20 fold the stoichiometric amount of the contained
Am) at pH = 1.5. The resulting precipitate when settled is redissolved in concentrated nitric acid, and is reprecipitated with
ammonium oxalate at pH = 1. Stirring the settled oxalate twice
with 0.2 $\underline{M}$ ammonium oxalate solution at a pH of about 10 reduces
the metallic impurities (especially Ni, Fe and Zn) so that > 99%
pure AmO_2 is obtained after calcination.

It is noteworthy that this process does not create any
additional solid waste, as all constituents of the waste solution
can either be recovered by destillation, or chemically decomposed
at relatively moderate temperatures. Corrosion attack is small
because no halogen compounds are involved, and the only potentially hazardous material is the ion exchange resin. When operated
at room temperature, there are no detectable signs of decomposition within one to two month's residence time. The spent resin
can be stored safely in alkaline media before incorporation into
concrete.

This relatively simple process has operated successfully
during 6 years with a total throughput of about 500 kg Pu and about
3 kg Am.

Pu/Am Separation by Extraction Chromatography

This study was carried out in order to evaluate the applica-
bility of extraction chromatography, with TBP as the extracting
agent, instead of anion exchange for efficient purification of
plutonium from [241]Am (18,19). The resin used was Levextrel-TBP, a
product of Bayer AG, Leverkusen, Germany. The Levextrels are
styrene - divinylbenzene - based resins which are copolymerized

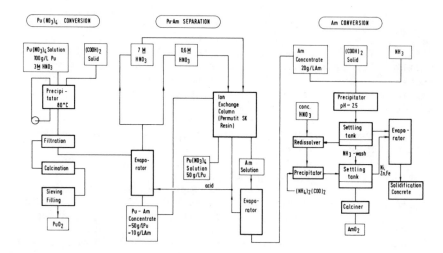

Figure 2.   Integrated flowsheet used at the ALKEM plant for plutonium nitrate conversion combined with ^{241}Am separation and conversion

in the presence of the respective extractant, e.g. TBP(20). These
resins offer the advantage that the extracting agent is better
fixed in the matrix material than is the case with extractants
simply sorbed on usual carrier materials; this is of particular
importance when technical-scale applications of extraction chromatography are envisaged.

Solid PuO_2 is the preferred form for intermediate storage
of plutonium . For dissolution of PuO_2 in nitric acid, fluoride
ion must be added as a catalyst. The distribution coefficients of
Pu(IV) between Levextrel-TBP and nitric acid(18) are considerably
reduced when F^- ion is present. This effect can be counteracted by
addition of Al^{3+} ion for complexation of F^-; an $Al^{3+} : F^-$ ratio
of 1 is sufficient for a satisfactory sorption of Pu(19). Maximum
loadings of 140 g/L resin have been obtained.

Laboratory-scale column experiments, using two resin columns
in series, were performed with feed solutions containing 25 g/L Pu,
90 mg/L Am, 6 $\underline{M}$ HNO_3, 0 to 0.55 $\underline{M}$ F^-, and 0 to 0.19 $\underline{M}$ $Al(NO_3)_3$.
With a flow rate of 5 mL/cm^2 . min, loadings of 100 to 120 g Pu/L
resin were obtained on the first column. The effluent from the
first column contained 11 to 22% of the Pu while that from the 2nd
column contained 0.02 to 0.9% of the Pu. Washing with 5 $\underline{M}$ HNO_3/
0.1 $\underline{M}$ $Al(NO_3)_3$ solution reduced the fluoride concentration to the
specified value (< 100 ppm). Elution of the 1st column with 3 bed
volumes of 0.3 to 0.5 $\underline{M}$ HNO_3 removed 69 to 87% of the Pu from the
column, with Am contents consistently lower than the specified
value of 100 ppm. Decontamination factors up to 350 for Am^{3+} and
up to 50 for F^- were measured.

For technical applications, knowledge of the irradiation be-
haviour of the Levextrel-TBP resin is important. A detailed study
carried out at the Radiochemistry Institute of the Technical University, Munich(21,22), showed that with gamma irradiation the
formation rate of dibutyl phosphoric acid (HDBP) and of "non-remo-
vable" acidic radiolysis products ("do-bads") is 2 to 5 times
lower with Levextrel-TBP resin than with pure TBP; the effect is
attributed to the "scavanger" action of the aromatic groups in the
matrix material. In summary, a high radiation resistance of the
resin has become evident.

A reference flowsheet for technical-scale operation of this
process, using three columns in series, has been proposed(19). In
Step 1 (loading), the feed solution (ca. 25 g/L Pu(IV), 3 to 6 $\underline{M}$
HNO_3, traces of Am^{3+} and F^-, with $Al(NO_3)_3$ added to Al : F = 3) is
fed at room temperature to the bottom of Col. 1 with a flow-rate
of < 3 mL/cm^2. min. The feed flow is stopped when the Pu concentration in the effluent from Col. 1 becomes > 70% of that of the
feed; a loading of 120 to 140 g Pu/L resin is obtained under this
condition. In Step 2 (scrub), 3 to 5 bed volumes of 5 $\underline{M}$ HNO_3 is
fed with a flow rate < 3 mL/cm^2 ·min to the bottom of Col. 1, with
Cols. 2 and 3 still in series. In Step 3 (elution), Col. 1 is dis-
connected from Cols. 2 and 3, and > 90% of the plutonium is eluted
from Col. 1 with 3 to 4 bed volumes of 0.3 $\underline{M}$ HNO_3 (50°C) in down-

flow dirction, with a flow-rate < 1 mL/cm^2 · min, to give a concentrated Pu procuct. The column is washed free from Pu with
additional O.3 $\underline{M}$ HNO$_3$; the tailings are recycled and combined with
the feed to the next cycle in which Col. 2 becomes the loaded
column, etc.

Processing of Neutron-irradiated ^{241}Am Targets

In the scope of the PACT project, a program was followed on
the production of medical-grade ^{238}Pu by neutron irradiation of
^{241}Am and alpha decay of the ^{242}Cm produced. This route would
offer a ^{238}Pu product which is essentially free from ^{236}Pu, and is
thus suitable for large medical power sources, e.g., for an arti-
ficial heart. The process which was studied for the chemical processing of irradiated AmO$_2$ - Al cermet targets (23) has recently
been reviewed(1); it consisted of the following steps:

(a) Aluminum is dissolved with 8 $\underline{M}$ NaOH.
(b) The NaOH solution is filtered off, and the AmO$_2$ residue is
    dissolved with 9 $\underline{M}$ nitric acid.
(c) Pu is adjusted to Pu(IV) and sorbed on Dowex 1X4 (< 400 mesh)
    resin on a high-pressure ion-exchange column. The column is
    washed with 7 $\underline{M}$ nitric acid, and the Pu is eluted with O.5 $\underline{M}$
    nitric acid.
(d) Pu is further purified by a second high-pressure anion ex-
    change cycle.
(e) For recovery of ^{241}Am the effluent of the first anion exchange
    cycle is denitrated by formic acid to O.5 $\underline{M}$ hydrogen ion(23).
(f) Am is sorbed together with rare earths (R.E.) and residual
    Cm on a high-pressure cation exchange column using AG 50X12
    resin (21 to 29 μm particle size). The adsorption column is
    washed free from other fission products with O.5 $\underline{M}$ NH$_4$NO$_3$
    solution.
(g) Am is separated from Cm and R.E.'s by chromatographic elution
    with O.5 $\underline{M}$ α-hydroxy isobutyric acid (pH = 3.45) through a
    high-pressure separation column loaded with AG 50X12 resin
    (21 to 29 μm).

Laboratory-scale tests with single irradiated AmO$_2$-Al cermet
pellets showed that the dissolution time of the aluminum matrix
must be kept to a minimum because ^{238}Pu losses increased severely
with increasing contact time of the concentrated NaOH. Proper adjustment of the plutonium valency was important to minimize ^{238}Pu
losses in the anion exchange separation; losses increased with
increasing ^{242}Cm concentration and, hence, alpha irradiation dose.
Treatment of the feed solution with hydrogen peroxide followed by
boiling for 1 h and immediate processing through the anion exchange column kept the ^{238}Pu losses down to about 5%. 90 to 98%
^{241}Am and 85 to 95% ^{242}Cm were recovered in the high-pressure
cation exchange step, with DF's of 100 to 300 for ^{95}Zr - ^{95}Nb,

> 200 for ^{103}Ru - ^{106}Ru and > 10000 for other fission products.

References

1. Koch, G.; "Recovery of by-product actinides from power reactor fuels and production of heat-source isotopes", ACS Sympos. Ser. No. 117 (1980), p. 411.

2. Ochsenfeld, W.; Baumgärtner, F.; Bauder, U.; Bleyl, H.J.; Ertel, D.; Koch, G.; Proc. Internat. Solv. Extract. Conf. ISEC 1977, vol. 2, p. 605; German Report KFK-2558 (1977).

3. Kolarik, Z.; Ochsenfeld, W.; KFK-Nachr. 11 (1979) No. 3,34

4. Koch, G.; German report KFK-1656 (1972) p. 1-10.

5. Koch, G.; Kolarik, Z.; Haug, H.; Hild, W.; Drobnik, S.; German report KFK-1651 (1972).

6. Koch, G.; Kolarik, Z.; Haug, H.; Radiochimiya (USSR) 17 (1975) 601; J. Inorg. Nucl. Chem., Suppl. 1976, 165.

7. Wheelwright, E.J.; Roberts, F.P.; Bray, L.A.; USA report BNWL-SA-1492 (1968).

8. Wheelwright, E.J.; Roberts, F.P.; USA report BNWL-1072 (1969).

9. Campbell, D.O.; Ind. Eng. Chem. Process Design Develop. 9 (1970) 95.

10. Hale, W.H.; Lowe, J.T.; Inorg. Nucl. Chem. Letters 5 (1969) 363.

11. Lowe, J.T.; Hale, W.H.; Hallmann, D.F.; Ind. Eng. Chem. Process Design Develop. 10 (1971) 131.

12. Burney, G.A.; Nucl. App. 4 (1968) 217.

13. Vaughen, V.C.A.; "Recovery of Americium and Curium", in: Koch, G. (ed.), "Transuranium Elements", Part A1 II, System No. 71 of "Gmelin Handbook of Inorganic Chemistry", Supplement Vol. 7b, p. 315-326, Springer, Berlin-Heidelberg - New York 1974.

14. Schulz, W.W.; "The Chemistry of Americium", ERDA Crit. Rev. Ser., TID 26971 (1976).

15. Koch, G.; Schön, J.; German report KFK-783 (1968).

16. Scheffler, K.; Kuhn, K.D.; Koch, G.; Schön, J.; Reaktortagung, Berlin 1970, Proceedings p. 534.

17. Schneider, V.; Koch, K.H. (ALKEM GmbH, Hanau); unpublished.

18. Ochsenfeld, W.; Schön, J.; Smits, D.; Tullius, E.;
     Kerntechnik 18 (1976) 258. Ochsenfeld, W.; Schön, J.;
     Reaktortagung, Mannheim 1977, Proceedings p. 381.

19. Eschrich, H.; Ochsenfeld, W.;
     Separation Science and Technology   15 (1980) 697.

20. Kroebel, R.; Meyer, A.; German patent application DE-OS
     2.162.951 (18 Dec. 1971).

21. de Waha, R.; Specht, S. (Technical University, Munich);
     unpublished.

22. Weh, R.; Specht, S. (Technical University, Munich);

23. Weinländer, W.; Bumiller, W.; German report KFK-1849 (1974)
     p. 54, and unpublished work reported in ref. (1).

RECEIVED December 30, 1980.

RECOVERY AND PURIFICATION
OF AMERICIUM-241

Status of Americium-241 Recovery at Rocky Flats Plant

JAMES B. KNIGHTON, P. G. HAGAN, J. D. NAVRATIL, and G. H. THOMPSON

Rockwell International, Box 464, Golden, CO 80401

^{241}Am grows into plutonium by the beta decay of ^{241}Pu. Americium is periodically removed from plutonium by a molten salt extraction process to lower the impurity content and to lower the gamma radiation associated with alpha decay of ^{241}Am to ^{237}Np. The extraction salt is an attractive source of ^{241}Am. At the Department of Energy's Rocky Flats Plant (RFP), the production scale recovery and purification of ^{241}Am from the extraction salts has involved aqueous ion exchange and precipitation processes.

Presently, about a kilogram per year of >95% AmO_2 (containing <1% individual contaminant elements) is produced and sent to the Department of Energy Isotope Pool at Oak Ridge National Laboratory. The americium is widely used in smoke detectors, oil well logging, thickness gauging, density, and radiographic measurements, and has many other uses because of its low energy gamma radiation.

Mullins, et al., separated americium from plutonium (during plutonium electrorefining) by partitioning americium between a molten salt containing plutonium +3 and molten plutonium metal (1). Knighton, et al., demonstrated that americium could be separated from plutonium by equilibrating molten chloride salts containing $MgCl_2$ with magnesium alloys, such as, Mg-Zn-Pu-Am. (2,3) Long, et al., investigated the distribution of americium between a molten NaCl-KCl salt containing 1.8 mole % $MgCl_2$ and molten plutonium metal (4). Production molten salt extraction processes subsequently were developed and implemented at Rocky Flats and at the Los Alamos Scientific Laboratory. At Rocky Flats, multikilogram quantities of plutonium metal are processed by the molten salt extraction process (5,6,7). Since the implementation of this process at Rocky Flats, improvements have been made to decrease the amount of salt requiring subsequent chemical processing for recovery of plutonium and americium. These improvements involve changing the salt composition by increasing the $MgCl_2$ content of the salt and by changing the mode of extraction from two stage crosscurrent to two stage countercurrent.

In spite of the improvements in the molten salt extraction

0097-6156/81/0161-0053$05.50/0

process, appreciable amounts of process residues remain; these residues contain kilogram quantities of plutonium as well as americium, both to be recovered by aqueous processes. The aqueous process used from 1968 to 1973 (the molten salt process was developed in 1967) comprised dissolution of the salt residues in water; precipitation with potassium hydroxide to remove the bulk of chloride; dissolution in 8M HNO_3; anion exchange recovery of plutonium; and purification of americium using the thiocyanate anion exchange process. (The thiocyanate anion exchange process had been developed in 1960 to process americium recovered from plutonium peroxide filtrate by hydroxide precipitation (8). Prior to development of the thiocyanate process, the americium hydroxide had been stored.) In 1973, the solvent used in salt dissolution was changed to 0.5M HCl, and in 1974, the potassium hydroxide precipitation step was replaced by a cation exchange process (9). The thiocyanate process was replaced by oxalate precipitation in 1975.

The process sequence currently used for waste salts (except those containing aluminum for which no process currently exists) is shown in Figure 1. The process includes (1) dilute hydrochloric acid dissolution of residues; (2) cation exchange to convert from the chloride to the nitrate system and to remove gross amounts of monovalent impurities; (3) anion exchange separation of plutonium; (4) oxalate precipitation of americium; and (5) calcination of the oxalate at 600°C to yield americium oxide.

The aqueous process portion of this paper describes attempts to improve the recovery of americium. The first part deals with modifications to the cation exchange step; the second describes development of a solvent extraction process that will recover americium from residues containing aluminum as well as other common impurities. (The anion exchange process cannot partition americium and aluminum.) Results of laboratory work are described.

This paper is presented in two parts: Part I, "Molten Salt Extraction of Americium From Molten Plutonium Metal" and Part II, "Aqueous Recovery of Americium From Extraction Salts."

PART I

MOLTEN SALT EXTRACTION OF AMERICIUM FROM MOLTEN PLUTONIUM METAL

Chemistry of Process

Americium is separated from plutonium by a liquid-liquid extraction process involving immiscible molten salt and molten plutonium metal phases. The molten salt extraction process is based upon equilibrium partitioning (by oxidation-reduction reactions) of americium and plutonium between the molten chloride salt and molten plutonium metal phases.

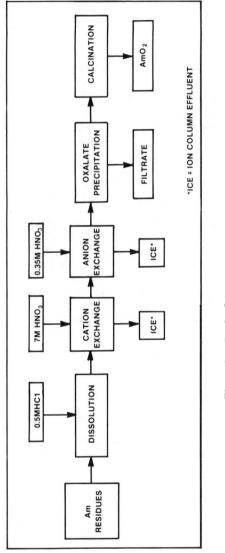

Figure 1. Rocky flats americium recovery process

The chemical basis for americium-plutonium separation is the free energies of formation (ΔG) of the metal chlorides: $MgCl_2$, $PuCl_3$, and $AmCl_3$.

Compound	ΔG_{1000} K kilocalories per gram atom of chlorine
$AmCl_3$	67 (See Reference 10)
$PuCl_3$	58.4 (See Reference 11)
$MgCl_2$	57.6 (See Reference 12)

From these values of the free energy of formation, it is reasonable to expect that both $PuCl_3$ and $MgCl_2$ would oxidize americium metal according to the reaction shown in Equations 1 and 2.

$$Am^\circ + PuCl_3 \rightarrow AmCl_3 + Pu^\circ \tag{1}$$

$$Am^\circ + 3/2MgCl_2 \rightarrow AmCl_3 + 3/2Mg^\circ \tag{2}$$

Magnesium chloride also oxidizes plutonium metal (Equation 3).

$$Pu^\circ + 3/2MgCl_2 \rightarrow PuCl_3 + 3/2Mg^\circ \tag{3}$$

At Rocky Flats, the metal composition is held essentially constant because the americium content (200 to 2000 ppm) and the amount of magnesium produced by Equations 2 and 3 are small and purified plutonium metal is the metal extraction product. Variables that can be manipulated and that influence the value of the distribution coefficient (Kd) are the salt composition and the temperature.

The distribution coefficient is defined as:

$$Kd = \frac{wt\ \%\ of\ solute\ in\ salt}{wt\ \%\ of\ solute\ in\ metal} \tag{4}$$

Extraction Theory

The transfer of a solute between immiscible liquid phases in liquid-liquid extraction is quantified by the following extraction factor relationship:

$$\alpha = Kd \cdot (s/m) \cdot F \cdot \beta \tag{5}$$

where

$$\alpha = \text{extraction factor} = \frac{wt\ of\ solute\ in\ salt\ phase}{wt\ of\ solute\ in\ metal\ phase}$$

Kd = distribution coefficient

s/m = salt-to-metal weight ratio

F = fraction of equilibrium

β = side reaction effects

The distribution coefficient (Kd) is dependent upon temperature, metal composition, salt composition, and solute concentration. The salt-to-metal ratio (s/m) is dependent upon the weights of the salt and metal feed. The fraction of equilibrium (F) is dependent upon the time and degree of mixing, and the side reaction term (β) is dependent upon the amount and kind of salt and metal insoluble impurities present in the system.

Distribution coefficients calculated from extraction data are influenced by (1) the value used for the salt-to-metal ratio, (2) the fraction of equilibrium obtained during the extraction, and (3) the extent of side reactions. Distribution coefficients calculated from extraction data are defined in this study as apparent distribution coefficients (Kd')

$$Kd' = Kd \bullet F \bullet \beta \qquad (6)$$

and are the product of three terms: the true distribution coefficient term (Kd), the fraction of equilibrium term (F), and the side reaction term (β). At equilibrium (when F = 1) and in the absence of side reactions (when β = 1), the value of Kd' is equal to Kd. By using the apparent distribution coefficient (Kd') rather than the true distribution coefficient (Kd), Equation 6 reduces to

$$\alpha = Kd' \bullet (s/m) \qquad (7)$$

From Equation 7, it follows that an infinite number of combinations exist for values of Kd' and s/m, the product of which equals any specific desired value of α.

Three modes of extraction (single-contact, two stage cross-current, and two stage countercurrent) have been used at Rocky Flats. In single-contact, one extraction stage is used. In two stage crosscurrent extraction, a batch of plutonium metal is contacted sequentially by two separate batches of new salt. In two stage countercurrent extraction, the salt and metal solvents move in opposite directions through the extraction stages.

From an operational viewpoint, a single-contact removal of americium is most desirable. To avoid secondary problems caused by the production of magnesium (by Equations 2 and 3) in excess of its solubility in plutonium, americium removals are limited to about 85% per extraction stage. Multiple-stage extractions consequently are used when 85% or greater removal of americium is required.

To lower americium content in the plutonium to acceptable levels, about 90% removal of americium is required. To obtain 90% removal with a two stage extraction, a separation of 68.4% is required in each stage. Magnesium generation is not a problem.

Crosscurrent extraction was used initially at Rocky Flats. This mode of extraction has the following disadvantages: (1) the solvent capacity of the salt is not used effectively, (2) plutonium loss to the salt is high because two salts must be equilibrated with the plutonium metal, and (3) a large amount of salt must be processed for recovery of americium and plutonium.

Countercurrent extraction is favored over crosscurrent extraction at a fixed salt composition for the following reasons: (1) less salt is required to obtain the same separation, (2) plutonium loss to the salt is lower, (3) less magnesium metal by-product is generated, and (4) salt feed to the salt recovery line is cut in half.

Table I gives the values of the extraction factor (α) required for americium removals ranging from 80 to 99%. These required values of α are given for the three modes of extraction (single-contact, two stage crosscurrent, and two stage countercurrent).

TABLE I.

Values of the Extraction Factor (α)

Percent Removal	Single-Contact	Countercurrent 2 Stages	Crosscurrent 2 Stages
80	4.000	1.562	1.236
85	5.667	1.933	1.582
90	9.000	2.540	2.162
91	10.111	2.719	2.333
92	11.500	2.928	2.536
93	13.286	3.179	2.779
94	15.667	3.489	3.083
95	19.000	3.887	3.472
96	24.000	4.425	4.000
97	32.333	5.208	4.773
98	49.000	6.519	6.071
99	99.000	9.460	9.000

The values of α required for a specific americium removal are peculiar to the extraction mode and the number of extraction stages.

The remaining terms in the extraction factor relationship (Kd, s/m, F, and β) and the interrelationship of the Kd' and s/m terms are discussed in the following sections of this paper.

Distribution Coefficient Term (Kd)

In the molten salt extraction process, the variables that control the values of the americium and plutonium distribution coefficients are temperature, metal composition, salt composition, and total americium. To minimize the variables, the extractions are conducted at a fixed temperature of about 750°C. Slight changes of magnesium content in the metal have a negligible effect upon the value of the americium and plutonium distribution coefficients. The effect of americium concentration in the metal upon the value of the americium distribution coefficient is believed negligible over the concentration range of 200 to 2000 ppm. Salt composition is therefore the variable with the major effect upon values of the americium and plutonium distribution coefficients. Since $MgCl_2$ is the oxidizing agent for americium and plutonium, values of the americium and plutonium distribution coefficients are expressed as functions of $MgCl_2$ content in the salt.

I. Johnson, at Argonne National Laboratory, derived a relationship for estimating the value of distribution coefficients for a solute partitioning between a salt of varying $MgCl_2$ content and a molten metal of fixed composition (13). This relationship is defined as:

$$D = C \ (X \ MgCl_2)^{3/2} \qquad (8)$$

where

$$D = \text{distribution coefficient} = \frac{\text{mole \% solute in salt}}{\text{atom \% solute in metal}}$$

C = constant

$X \ MgCl_2$ = mole fraction $MgCl_2$ in salt

For convenience, Equation 8 may also be expressed in terms of Kd rather than D. The values of the constant, however, are obviously different.

$$Kd = C' \ (X \ MgCl_2)^{3/2} \qquad (9)$$

The distribution coefficients for americium and plutonium are estimated by using the appropriate constants in Equation 9 as shown in Equations 10 and 11.

$$KdAm = 273.7 \ (X \ MgCl_2)^{3/2} \qquad (10)$$

$$KdPu = 0.692 \ (X \ MgCl_2)^{3/2} \qquad (11)$$

From equations 10 and 11, values of the americium and plutonium distribution coefficient can be estimated for $MgCl_2$ contents in the salt ranging from 0.02 to 1.0 mole fraction (Table II).

TABLE II.

Estimated Values of Americium and Plutonium Distribution
Coefficients for NaCl-KCl-MgCl$_2$ Salt and Plutonium
Metal System at 750°C

MgCl$_2$ Mole Fraction	Estimated KdAm	Estimated KdPu
0.02	0.77	0.002
0.05	3.1	0.0085
0.10	8.7	0.022
0.15	15.9	0.040
0.20	24.5	0.063
0.30	45.0	0.115
0.40	69.2	0.175
0.60	127.3	0.325
0.70	160.4	0.405
1.00	273.4	0.693

Salt-to-Metal Ratio Term (s/m)

The salt-to-metal term (s/m) is the weight ratio of the
liquid salt and metal phases present at equilibrium. The actual
weights of the liquid salt and metal present at equilibrium may
be estimated from the feed weights by factoring in the weight
changes caused by (1) transfer of plutonium and americium from
the metal to the salt, (2) transfer of magnesium from the salt to
the metal, (3) salt take-up of plutonium insoluble impurities
associated with the plutonium metal feed, (4) build-up and
release of salt and metal on the crucible and stirrer, and (5)
evaporation of volatiles, such as Mg, from the metal.

Because of the above uncertainties in estimating actual
weights of liquid salt and molten plutonium at equilibrium, it is
more practical (although not rigorous) to base the salt-to-metal
ratio on the weights of the salt and metal fed to the extraction
rather than on the estimated weights of the salt and metal at
equilibrium. This puts a low bias on the value of the
salt-to-metal ratio and a high bias on the value of the apparent
distribution coefficient.

Mixing Term (F)

Two unit operations are used in the equilibration of the
salt and metal phases: (1) intermixing of salt and metal, and
(2) disengagement of salt and metal. Because this is a batch
extraction, both operations (intermixing and disengagement of
phases) occur sequentially in the same vessel. For practical
operation of the molten salt extraction process, attainment of
equilibrium or near-equilibrium conditions (when the value of F
approaches 1) in a relatively short period of time is essential.

Complete disengagement of phases in a relatively short period of time also is essential.
To provide continuous intermixing of the light and heavy phases in an unbaffled crucible, a reverse-motion mode of mixing was developed. This reverse-motion mixing was obtained by repeating the following sequence of events: 2.5 sec clockwise stirrer rotation, 0.5 sec stop, 2.5 sec counterclockwise stirrer rotation, and 0.5 sec stop. The above times for clockwise and counterclockwise mixing and for the stop periods have not been optimized; however, they are adequate for operations at Rocky Flats. By frequently reversing the direction of mixing, the stirrer blade also serves as a baffle to intermix the swirling light and heavy phases.
It is believed that equilibrium conditions are closely approached with the reverse-motion mode of mixing (F ≃ 1).

Side Reaction Term (β)

Salt and metal insoluble impurities, such as PuO_2, associated with plutonium metal are taken up by the salt in Stage 1. Stage 2 is essentially free of these impurities. Strickland, et al. (14), reported that plutonium oxide extracts americium from molten plutonium metal in a molten salt media. Because these salt and metal insoluble impurities are present in sizable amounts only in Stage 1, the side reaction between americium and these impurities occurs only in Stage 1. The side reaction term (β) is introduced to quantify the side reaction caused by the presence of impurities such as in Stage 1.

The ratio of Equation 6 for Stages 1 and 2 gives

$$\frac{Stage\ 1}{Stage\ 2} = \frac{Kd'_1}{Kd'_2} = \frac{Kd_1 \cdot F_1 \cdot \beta_1}{Kd_2 \cdot F_2 \cdot \beta_2} \tag{12}$$

Because of the absence of salt and metal insoluble impurities in Stage 2, the value of the side reaction term for Stage 2 is $\beta_2 = 1$. The value of the true distribution coefficient (Kd) is assumed to be the same for both stages. The effect of americium concentration (200 to 2000 ppm) in the metal upon the value of Kd is assumed to be negligible over the above concentration range. Because the same mode and time of mixing are used in Stages 1 and 2, the value of F (fraction of equilibrium) is assumed to be the same for both stages.
From the above assumptions ($Kd_1 = Kd_2$, $F_1 = F_2$, and $\beta_2 = 1$), Equation 12 reduces to Equation 13 and provides a method for estimating the value of the side reaction term (β_1) for Stage 1.

$$\beta_1 = \frac{Kd'_1}{Kd'_2} \tag{13}$$

Process Optimization

An optimum molten salt extraction process at Rocky Flats would use the minimum amount of salt required to obtain (1) a desired removal of americium, (2) a minimum transfer of plutonium to the salt, and (3) a minimum take-up of magnesium by the plutonium metal product. The product salt must be compatible with subsequent chemical processes for the recovery of americium and plutonium contained in the salt. To minimize the number of glove-box operations, time in the gloves, and operator radiation exposure, the operations must be simple and easy to conduct. By using the minimum amount of salt feed, a minimum amount of waste will be generated that ultimately must be sent to long-term storage.

To optimize the molten salt extraction process, the values of the terms in the extraction factor relationship

$$\alpha = Kd' \cdot (s/m) \tag{7}$$

and the interrelationship of these terms must be known.

The numerical values of the extraction factor (α) are set by selecting the desired separation, the mode of extraction, and the number of extraction stages. For example, the value of α required for 90% americium removal by two stage countercurrent extraction is $\alpha = 2.54$ (see Table I). As shown previously, the value of the americium distribution coefficient is a function of the salt composition; i.e., the $MgCl_2$ content of the salt and the composition of the diluent salt system. The value of the salt-to-metal ratio is set by the weight of salt and metal fed to the extraction.

When the values of any two of the three terms in Equation 11 are known, the value of the third term may be calculated. From Equation 11, it is noted that there are infinite combinations of values for Kd' and s/m, the product of which equals any specified value of α. As the value of Kd' becomes large, the corresponding value of s/m becomes small when the value of α is held constant.

Figure 2 gives the kg salt per kg Pu (or s/m) and the corresponding $MgCl_2$ content in the $NaCl-KCl-MgCl_2$ salt system. These data are for americium removals ranging between 80 and 99% using the two stage countercurrent mode of extraction. Figure 2 provides the basis for optimizing the process. The minimum amount of salt that can be physically handled in the extraction is determined and is represented by a horizontal line. The optimum $MgCl_2$ content in the salt occurs where this horizontal line intersects the line representing the desired removal of americium.

In the actual operation of the extraction process, the minimum amount of salt is about 0.05 kg salt per kg Pu. This amount of salt is barely sufficient to cover the molten plutonium. Even with this small amount of salt, molten plutonium is exposed to the cell atmosphere during mixing. Impurities in

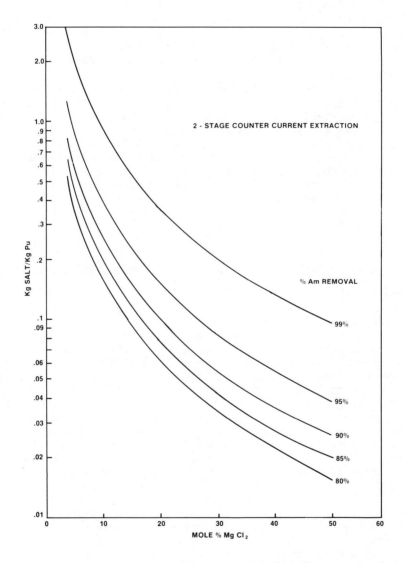

Figure 2. Relationship between salt-to-metal ratio and $MgCl_2$ content in salt for selected removals of americium from plutonium $NaCl$–KCl–$MgCl_2$ salt

the cell atmosphere may react with the plutonium metal to form
compounds that are taken up by the salt. High plutonium losses
to the salt are undesirable.

The intersection of the 0.05 kg salt per kg Pu line with the
curve for 90% americium occurs at 32 mole % $MgCl_2$. For
convenience purposes only, a 30 mole % $MgCl_2$ salt combined with
a salt-to-metal ratio of 0.056 is near optimum.

Table III compares the previous production process with an
optimum process.

In the previous production process, an americium removal of
90% was obtained using a salt containing 5.74 mole % $MgCl_2$ at a
salt-to-metal ratio (s/m) of 0.557. By comparison, the optimum
process gives the same americium removal (90%) with a 30 mole %
$MgCl_2$ salt at a salt-to-metal ratio of 0.0564. The amount of
salt is decreased by an order of magnitude with the optimum
process, and the americium concentration in the extraction salt
is increased by an order of magnitude. Optimum extraction
conditions (minimum amount of high $MgCl_2$ salt) were implemented
in the production operation of the molten salt extraction process.

PART II

AQUEOUS RECOVERY OF AMERICIUM FROM EXTRACTION SALTS

Cation Exchange

In the cation exchange process, plutonium and americium are
cosorbed on the resin with mono- and divalent cations from the
molten salt extraction (MSE) process (15). After actinide
breakthrough, the column is washed with 0.35M HNO_3 to remove
nonadsorbed ions remaining, then eluted with 7M HNO_3. Plutonium
is oxidized to Pu(IV) and forms the hexanitrato complex
$Pu(NO_3)_6^{2-}$. Although this anionic complex is not adsorbed on the
resin, it still does not elute very rapidly. This is attributed
to both the difficulty the bulky complex has in diffusing through
the resin matrix, and the high selectivity of the resin for
Pu(IV). The process was made easier by using gel-type resin of
lower cross linkage and macroporous resin, as reported earlier
(16).

A continuing problem with the cation exchange process as
used in production operations is that it has not been
sufficiently selective and therefore allows considerable
carryover of the MSE salt constituents and impurities with the
plutonium and americium. This isn't serious with plutonium since
plutonium can be subsequently purified by anion exchange. For
americium, however, the subsequent recovery process is oxalate
precipitation which is less selective and carries some of the
impurities into the final product.

Originally, selective separation by cation exchange was
expected to result because of large differences in the

TABLE III

Comparison of Previous Production and Optimum MSE Processes

	Previous Production Process	Optimum Process
Extraction mode	countercurrent	countercurrent
Extraction stages	2	2
Pu recovery	97%	97%
Am removal	90%	90%
Extraction factor (α)	2.54	2.54
Weight Pu metal feed	~ 2.2 kg	~ 2.2 kg
Weight salt feed	~ 1.225 kg	~ 0.123 kg
Salt-to-metal ratio (s/m)	~ 0.557[a]	~ 0.0564[a]
$MgCl_2$ in salt	5.74 mole %[b]	30.0 mole %[c]
Am distribution coefficient (Kd')	4.56[d]	45.0

[a] Slight adjustments periodically are made on the s/m value because the quality of Pu metal and Am content varies.
[b] 5.74 mole % $MgCl_2$, 47.13 mole % NaCl, and 47.13 mole % KCl.
[c] 30 mole % $MgCl_2$, 35 mole % NaCl, and 35 mole % KCl.
[d] Kd' varies depending upon the amount of PuO_2 present during the extraction. Value of true distribution coefficient is Kd = 3.76.

distribution values of the actinides versus those of the MSE
waste salt constituents. The problem with this assumption is the
dependency of the distribution value (or selectivity coefficient)
upon factors such as temperature, pressure, complexing ions, and
ionic strength. The latter is especially important in separating
ions of different valences.

Other consideration for selective separation are related to
the basic principles of ion exchange. For example, cation
displacement should proceed until the column is loaded to
capacity with plutonium and americium. If this procedure is not
employed, the excess resin simply sorbs the displaced cations
(e.g., Mg, Na, K, Ca, etc.) and they are rejoined with the
actinides during the elution cycle.

In addition to loading the column to breakthrough, there is
an advantage in using a column washing cycle. Cations can be
selectively displaced by washing with an ion that will not
further contaminate the product, i.e., hydrogen ion; the rate of
displacement depends upon the acid concentration.

An investigation was therefore made on the effects of ionic
strength and column washing procedures. The results were
analyzed and specific recommendations were made for operational
changes; results and recommendations are given in this report.

Experimental

A synthetic MSE waste salt was prepared with a composition
that has been projected as a new waste salt product from
pyrochemical processing: 2.3 wt% Am, 22.6 wt% Pu, 10.3 wt% KCl,
21.4 wt% $MgCl_2$, and 43.1 wt% $CaCl_2$. The most significant change
in this salt from that currently produced is the substitution of
$CaCl_2$ for NaCl; calcium is much more difficult to remove than
sodium. (It also coprecipitates with americium during oxalate
precipitation and is therefore a serious contaminant.) In
addition, the salt was also made 0.3 wt% in Pb, a common
impurity in MSE waste salts.

A waste salt feed solution was then prepared by dissolving
50 g of the salt in 500 mL of 0.5N HCl. This yielded a feed
solution containing 2.3 g/L Am; 22.6 g/L Pu; 5.4 g/L K;
15.6 g/L Ca; 5.5 g/L Mg; 0.3 g/L Pb; and 48.3 g/L Cl. The total
molar concentration was about 2.7.

The ion exchange column was 1.12 cm inside diameter by
30.5 cm (column capacity 30 cm^3). The column was filled with
Dowex 50W-X8 (50-100 mesh) cation exchange resin, the resin
currently used in production operations. A peristaltic pump was
used to pump solutions through the resin bed. Column loading and
washing were done in an up-flow mode, and elution was done in a
down-flow mode. The flow rate was 3 mL/min.

To determine the effects of ionic strength, 32 mL of feed
was passed through the column in one of three different ways:
(1) without dilution, (2) diluted by a factor of five with

0.5N̲ HCl, or (3) diluted by a factor of 20 with 0.5N̲ HCl. The length of the plutonium band on the column was measured and the effluent solutions were analyzed for cation concentration. Plutonium in the effluent solutions was determined by radiometric counting, the americium by gamma spectroscopy, and the remaining cations by atomic absorption.

After the feed solutions were passed onto the resin, the column was washed with four column volumes of 0.35N̲ HNO₃ to remove residual chloride ions. The column was then washed with four column volumes of one of the following acid solutions: 0.35, 1.0, 1.5, or 2N̲ HNO₃. This second wash was to determine if remaining cations could be selectively removed from the resin. Following the column washing steps, the plutonium and americium were stripped from the column with 7.2N̲ HNO₃. The effluent solutions from the washing and elution cycles were analyzed, as before, for plutonium, americium, and salt constituent concentration.

Results and Discussion

To determine column loading characteristics as a function of ionic strength, MSE salt solutions of different molar strengths were pumped onto a column of cation exchange resin and the distance the plutonium migrated through the column was measured. As shown in Table IV, when the salt solution is passed onto the resin in a concentration strength similar to production operations (2.7M̲), the plutonium band migrated about 25 cm or 86% of the column length. In contrast, when the feed was diluted by a factor of five (to 0.9M) using 0.5N̲ HCl, the plutonium band migrated only 13 cm or 45% of the column length. Increasing the dilution to a factor of 20 appeared to further decrease migration; however, the decrease is only slight.

TABLE IV.

Migration of Plutonium Through Dowex 50W-X8
Cation Exchange Resin

Test No.	Dilution Factor	Pu Migration, cm	% Column Loaded
1	0	26.0	86.7
2	0	24.5	85.8
3	5	14.5	48.3
4	5	12.5	41.7
5	20	12.0	40.0

This study shows that by diluting the salt solution, the selectivity for plutonium and americium is greatly increased; whereas   in undiluted solutions the actinides tend to migrate through the resin with the other salt constituents. Selectivity between different valencies is a function of the ionic strength, and the selectivity of the resin for the trivalent ion over the divalent ion (such as Pu^{3+} over Ca^{2+}) is inversely related to the total concentration of the solution (16). Furthermore, the selectivity of the resin for the trivalent ion over the monovalent ion (such as Pu^{3+} over K^+ or Na^+) is inversely related to the square of the total concentration.

Column Washing

After the salt solutions had been passed through the column and the migration of plutonium measured, the resin was washed with four column volumes of $0.35\underline{N}$ HNO_3. This was done to remove any residual chloride ions. A study was then made to determine if remaining cationic impurities could be selectively displaced using low concentrations of nitric acid. The levels selected were 0.35, 1.0, 1.5, and $2\underline{N}$. As shown in Table V, when the resin was washed for the second time with $0.35\underline{N}$ HNO_3 additional quantities of impurities are removed: 14% of the Ca, 44% of the Mg, 11% of the Pb, and 90% of the K. The displacement, however, can be greatly increased by using slightly stronger nitric acid. For example, $1.5\underline{N}$ nitric acid displaced 90% of the Ca, 98% of the Mg, 96% of the Pb, and greater than 99% of the K.

TABLE V.

Percent Impurity Removed As A Function Of
Nitric Acid Concentration [a,b]

Element Removed	Nitric Acid Concentration,N			
	0.35	1	1.5	2.0
Pu	<0.1	<0.1	0.4	7.0
Am	<0.1	<0.1	<0.1	2.2
Ca	14.4	61.0	89.6	96.8
Mg	43.6	90.1	98.5	>99.9
Pb	11.5	66.6	95.5	94.5
K	90.0	90.5	>99.9	>99.9

[a] These comparative washing values are only from column studies with feed diluted by a factor of five.
[b] Results are based on single determinations.

Washing with nitric acid does cause some slight displacement of plutonium, especially at the 2N level. Since this solution is free of chloride ion and has few impurities, it could be recycled through anion exchange for secondary (low-level) recovery.

Table VI shows the step or cycle (i.e., loading, first wash, second wash, or elution) during which the cations were separated from the resin. The results tabulated are only with those solutions diluted by a factor of five. As shown, the plutonium and americium were not eluted by washing with nitric acid until the wash solution approached 2N. The other ions were removed to different degrees in the loading and first washing cycle, and then greatly influenced by the change in the acid concentration of the second wash. (We cannot explain the large difference in effluent composition for the loading and first washing cycle operations; these operations were identical throughout.)

Of special interest was the removal of lead. Most lead was removed during the loading and first washing cycle, which indicates lead is being complexed by the chloride ion. This complex is currently being used to aid in the decontamination of lead in the oxalate precipitation process.

Conclusions

This work demonstrates that considerable improvement in the cation exchange process can be made by incorporating the following changes into the process procedures:

1) The feed solution should be diluted by a factor of at least five using 0.5N HCl.

2) Loading the column should proceed until the plutonium and americium near breakthrough; both migrate at about the same rate.

3) Column washing should be a two-step procedure. The first wash with 0.35N nitric acid is to remove the chloride ion, and the second wash with 1.5N nitric acid is to remove residual impurity cations.

The dilution of the feed stream will allow for selective displacement of the cations, and therefore, increase the capacity of ion exchange columns. Loading the columns to near breakthrough capacity will displace the impurity cations so that the subsequent steps of processing will be more effective.

Washing the column with 0.35N nitric acid first will not only remove the chloride ions, but also prevents the formation of hydrolytic or polymeric species of plutonium in comparison with the water wash formerly used. The second wash has been shown to effectively displace the impurity cations; however, some bleeding of plutonium and americium may occur and cause higher

TABLE VI.

Element Separation by Cation Exchange

Second Wash N HNO$_3$	Element	Percentage of Element in Ion Column Effluent			
		Loading	0.35N HNO$_3$ Wash	Second Wash	Elution
0.35	Pu	<0.1	<0.1	<0.1	>99.9
	Am	<0.1	<0.1	<0.1	>99.9
	Ca	<0.1	13.9	12.4	73.7
	Mg	9.8	66.6	10.3	13.3
	Pb	61.3	28.3	1.2	9.2
	K	16.0	65.0	17.1	1.9
1.0	Pu	<0.1	<0.1	<0.1	>99.9
	Am	<0.1	<0.1	<0.1	>99.9
	Ca	<0.1	2.0	59.8	38.2
	Mg	0.4	55.0	40.2	4.4
	Pb	16.6	46.0	24.9	12.5
	K	<0.2	74.7	22.9	2.4
1.5	Pu	<0.1	<0.1	0.4	99.6
	Am	<0.1	<0.1	<0.1	>99.9
	Ca	<0.1	10.7	80.0	9.3
	Mg	0.5	53.6	45.2	0.7
	Pb	<2.5	44.3	53.2	2.5
	K	14.1	55.1	30.8	<0.1
2.0	Pu	<0.1	<0.1	7.0	93.0
	Am	<0.1	<0.1	2.2	97.8
	Ca	<0.1	9.5	87.6	2.9
	Mg	<0.1	51.6	48.4	<0.1
	Pb	<2.5	29.7	66.4	3.9
	K	1.8	69.0	31.3	<0.1

than usual levels of the actinides in the effluent. If this happens, it is suggested that the effluent from the second wash be returned to secondary (low-level) recovery operations.

Future Work

Cation exchange work now in progress comprises evaluation of the optimized cation exchange process with the BioRad AG MP-50 macroporous resin that is now a replacement candidate for the Dowex 50W-X8 gel-type resin. Other work being considered includes use of HCl wash solutions prior to conversion to the nitrate system and possible use of chelating agents.

Additional work in progress includes optimization of parameters affecting the oxalate precipitation step; this includes determination of the chloride concentration required to solubilize lead; the oxalate ion concentration required for maximum americium recovery with minimum impurity precipitation; precipitate aging; and hydrogen ion concentrations that will minimize americium solubility yet maximize impurity solubilization.

Bidentate Extraction

Recovery of actinides at the RFP with an organic phosphorous bidentate extractant has been proposed. A conceptual production flow sheet is shown in Figure 3. The bidentate reagent, dihexyl-N, N-diethylcarbamoylmethylenephosphonate (DHDECMP), is especially attractive since it can recover actinides from MSE residues containing aluminum. The cation exchange process is unable to effect actinide purification when aluminum is present. (DHDECMP extracts actinides and lanthanides, but does not extract common RFP contaminants, e.g., aluminum. No lanthanides are used in process streams at RFP.)

The actinides are extracted from high acid (e.g., $7\underline{N}$ HNO_3) solutions and can be back-extracted with dilute acid. The method is therefore easily used with the column effluent from the anion exchange plutonium recovery step.

Two techniques appear to be useful for the bidentate extraction of actinides. The first is liquid-liquid solvent extraction, a method which has several advantages. Currently, however, the type of equipment needed (mixer-settlers, centrifugal contactors, etc.) is not available at RFP. We are better equipped to use a chromatographic column technique. This comprises sorbing the bidentate extractant on an inert solid support, loading ion exchange columns with the sorbent, then passing solutions through the columns.

Both liquid-liquid extraction and extraction chromatography have been tested. Laboratory and pilot scale recovery tests were done using extraction chromatography; results of these preliminary tests were described previously (15).

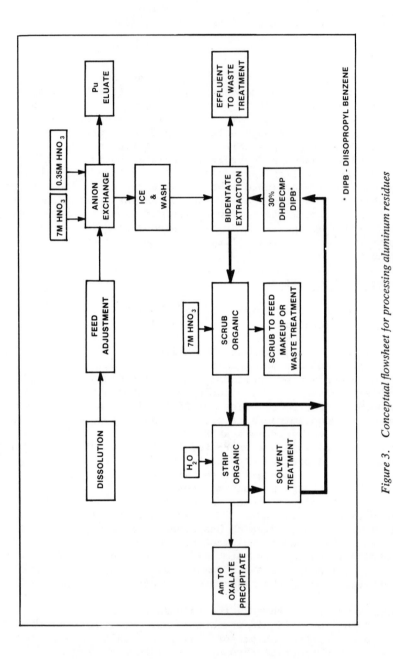

Figure 3. Conceptual flowsheet for processing aluminum residues

We are now involved in the design, fabrication, and installation of a pilot plant scale solvent extraction system. The glove box is in place. Design of the facility is essentially complete, as is the procurement of off-the-shelf equipment items. Fabrication and installation are in progress.

The extractors are York rotating disc contactors, 3-inch ID by 48-inch long. Three contactors are to be used, one each for extraction, scrubbing, and stripping. The system will be used to investigate the recovery of plutonium and americium from MSE waste salt solutions using 30% DHDECMP in diisopropylbenzene.

Pilot plant scale testing of the extraction chromatography process is currently in progress at Rocky Flats Plant. When adequate testing of both the solvent extraction and extraction chromatography methods has been accomplished, the methods will be compared to see which has the greatest promise for recovering the americium from special MSE residues.

Acknowledgment

This work was performed under a contract with the U. S. Department of Energy. Rockwell International Corporation and the United States Government expressly reserve the right to print, reprint, publish, copy, vend, translate, and use any or all material contained herein.

Literature Cited

1. Mullins, L.J.; Leary, J.A.; Morgan, A.E., U.S. AEC Report, LA-2666, (1962).
2. Knighton, J.B.; Steunenberg, R.K., U. S. Patent 3 147 109, (1964).
3. Knighton, J.B.; Steunenberg, R.K., J. Inorg. Nucl. Chem. (1965) 27, 1457.
4. Long, J.L., U. S. Patent 3 460 917, (1969).
5. Long, J.L.; Perry, C.C., in "Symposium on Reprocessing of Nuclear Fuels." P. Chiotti, Ed., Nuclear Metallurgy, Conference 690801, August 19; 15, p. 325
6. Knighton, J.B.; Long, J.L.; Franchini, R.C.; Auge, R.J.; Brown, J.C.; Meyer, F.G., U.S. AEC Report, RFP-1875, (1973).
7. Knighton, J.B.; Auge, R.G.; Berry, J.W.; Franchini, R.C., U. S. ERDA Report, RFP-2365, (1976).
8. Ryan, V.A.; Pringle, R.W., U.S. AEC Report RFP-130, (1960).
9. Proctor, S.G., U.S. ERDA Report RFP-2347, (1975).
10. Glassner, A., U.S. AEC Report, ANL 5750, undated.
11. Oetting, F.L., Chem. Rev. (1967) 67, 261.
12. Wicks, C.E.; Block, F.E., U.S. Bur. Mines Bull. 605 (1963).
13. Johnson, I., Argonne National Laboratory, personal communication, (1968).
14. Strickland, W.R.; Downing, W.E.; Brown, J.L.; Auge, R.G.; Rocky Flats Plant, unpublished data, (1969).

15. Navratil, J.D.; Martella, L.L.; Thompson, G.H., in
     "Actinide Separations," Navratil, J.D.; Schulz, W.W., Eds.,
     Symposium Series No. 117, American Chemical Society:
     Washington, D.C., (1980), p. 455.
16. "Dowex: Ion Exchange"; Dow Chemical Company: Midland,
     Michigan, (1964).

RECEIVED March 19, 1981.

Status of Americium-241 Recovery and Purification at the Los Alamos National Laboratory

HERMAN D. RAMSEY, DAVID G. CLIFTON, SIDNEY W. HAYTER, ROBERT A. PENNEMAN, and ELDON L. CHRISTENSEN

University of California, Los Alamos National Laboratory, Los Alamos, NM 87545

Separation of americium from plutonium began at the Los Alamos National Laboratory in late 1947 with the formation of a small group for that purpose headed by R. A. Penneman. The early investigations centered necessarily on the isolation of americium from plutonium which had sufficient irradiation and age that the beta decay of ^{241}Pu, $t_{\frac{1}{2}}$ = 14.4 yr., would yield appreciable ^{241}Am (150 - 200 mgs/Kg of Pu). A few kilograms of plutonium turnings were processed by Los Alamos yielding about a gram of americium. The plutonium had been through the $BiPO_4/LaF_3$ process and contained 100 times as much lanthanum as americium; this lanthanum naturally separated with the crude americium. Later, filtrates were processed from the plutonium peroxide process used in the production plant (Hanford, Washington). Plutonium that had been recovered by the Purex process yielded americium with less impurities to contend with.

Americium was isolated first from plutonium, then from lanthanum and other impurities, by a combination of precipitation, solvent extraction, and ion exchange processes. Parallel with the separation, a vigorous program of research began. Beginning in 1950, a series of publications (1-24) on americium put into the world literature much of the classic chemistry of americium, including discussion of the hexavalent state, the soluble tetravalent state, oxidation potentials, disproportionation, the crystal structure(s) of the metal, and many compounds of americium. In particular, use of peroxydisulfate or ozone to oxidize americium to the (V) or (VI) states still provides the basis for americium removal from other elements. Irradiation of americium, first at Chalk River (Ontario, Canada) and later at the Materials Testing Reactor (Idaho), yielded curium for study. Indeed, the oxidation of americium and its separation from curium provided the clue utilized by others in a patented process for separation of americium from the rare earths.

0097-6156/81/0161-0075$05.00/0
© 1981 American Chemical Society

Figure 1.   Plutonium facility at Los Alamos National Laboratory

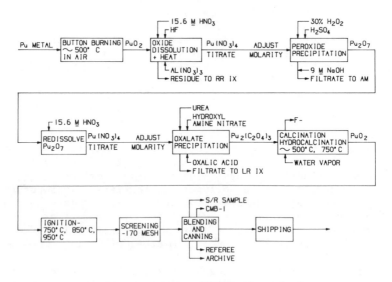

Figure 2.   Flowsheet for FFTF oxide production

Production.

At present, americium is separated and purified in Kg/yr
quantities at the new plutonium facility at Los Alamos (Figure
1). The feed for the americium production comes from a line
which produces high purity ceramic grade PuO_2 for the Fast Flux
Test Facility (FFTF) at Richland, Washington. The feed for
this FFTF PuO_2 is aged plutonium metal which contains sizable
amounts of ^{241}Am
    The flowsheet for the FFTF PuO_2 production is shown in
Figure 2. Briefly, the plutonium metal is converted to an impure oxide by burning the metal in air. This is followed by
dissolution of the impure oxide in a 15.6 $\underline{M}$ HNO_3 - 0.5 $\underline{M}$ HF
solution. The americium is separated from the plutonium by
precipitation of the plutonium as the peroxide. Americium does
not form an insoluble peroxide and stays in the filtrate with
other cationic impurities. The active peroxide filtrate is
slowly dripped into 9 $\underline{M}$ NaOH. The combination of strong alkali
and heat destroys the peroxides and precipitates the americium
as the hydroxide. Any residual plutonium in the filtrate,
along with other cations, is precipitated also as the hydroxide.
The flowsheet for the americium oxide production is shown in
Figure 3.
    Upon cooling, this slurry of hydroxides is transferred by
vacuum into glass tanks (Figure 4). These tanks are shielded
with teflon coated lead to minimize radiation exposure to personnel. The lead is teflon coated to prevent contamination with
lead in the americium product. The slurry of hydroxides is
gelatinous in nature and is difficult to filter. To improve
the filterability of these hydroxides, the slurry is allowed
to stand overnight and settle (Figure 5). The hydroxides seem
to aggregate upon standing. Also, the filtering process is
faster after settling as the clear supernant liquid above the
aggregated hydroxides can be decanted and passed through the
filter very fast. The remaining hydroxides are then slowly
filtered with vacuum onto filter paper in a 20 cm stainless
steel filter boat (Figure 6), rinsed with 0.1 $\underline{M}$ NaOH to remove
excess sodium salts, and redissolved in 15.6 $\underline{M}$ HNO_3 and sent
to the next step in the process. After analysis for Am and Pu,
the filtrate from this filtration is sent to waste. The filter
paper is rinsed well with water, dried, and incinerated. The
compositions of two representative batches of peroxide filtrates
are shown in Table I.
    The next step involves removal of residual plutonium from
the americium by ion exchange chromatography. The americium
nitrate solution is passed, after acidity adjustment to 7.5 $\underline{M}$,
through a 15 cm x 30 cm ion exchange column containing Dowex
1-X4, 50 - 100 mesh (Dow Chemical Company). Periodically, an
unknown gel is observed in the americium nitrate solution and

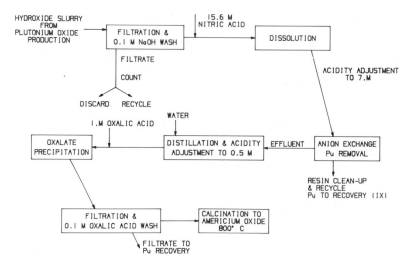

Figure 3.   Flowsheet for americium oxide production

Figure 4.   Americium hydroxide slurry holding tanks with Teflon-coated shielding

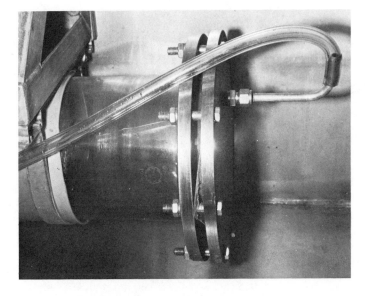

Figure 5.   Americium hydroxide slurry after standing 12 h

Figure 6.   Americium hydroxide in stainless steel filter boat

T A B L E   I

COMPOSITION OF PEROXIDE FILTRATES

The filtrates were approximately 0.03 $\underline{M}$ in F^- and 0.056 $\underline{M}$ in $SO_4^=$. These filtrates were dripped into approximately 80 moles of NaOH. Values are mg/L unless otherwise noted.

Batch No. / Element	1049	1127
Am	0.26 g/L	0.41 g/L
Pu	0.03 g/L	0.03 g/L
Ag	< 0.15	< 0.1 g/L
Al	500	200
B	25	4
Ba	0.5	0.8
Be	< 0.05	< 0.04
Bi	< 1.5	< 1
Ca	< 25	< 30
Cd	< 5	< 4
Co	< 1.5	< 1
Cu	5	1.6
Cr	5	4
Ga	< 0.5	< 0.4
Ge	< 0.5	< 0.4
Fe	15	16
In	< 1.5	< 1
K	< 50	< 40
Li	< 5	< 4
Mg	25	8
Mn	1	0.8
Mo	3	< 1
Na	75	20
Ni	10	2
Nb	< 5	< 4
P	< 5	< 4
Pb	2	4
Sb	< 1.5	< 1
Si	20	20
Sn	< 3	< 2
Sr	0.25	0.6
Ta	< 50	< 40
Ti	1.5	< 0.4
Tl	< 1.5	< 1
V	< 1.5	< 1
Zn	< 3	< 2
Zr	< 0.5	< 0.4

must be filtered off to prevent plugging of the ion exchange
column. The ion exchange column is gravity fed and is also
shielded with a layer of teflon coated lead. The usual problem
(gas generation from radiolysis of the solution) of running
concentrated americium solutions through ion exchange resins
is encountered. The ion exchange column is run on a batch
schedule. Concentrated americium solution is never allowed to
stand on the column. After a batch has passed through the column, the column void volume containing americium nitrate solution is replaced with an excess of 7 $\underline{M}$ nitric acid solution to
wash out all of the americium effluent. After several batches
of americium have been passed through the ion exchange column,
the anion resin becomes loaded with $Pu(NO_3)_6^=$ and must be eluted.
The acidity of the solution on the ion exchange column is lowered to approximately 1 $\underline{M}$, and the plutonium is reduced with a
hydroxylamine nitrate solution to Pu^{+3}. It is worthy to note
the violent reaction that occurs between hydroxylamine nitrate
and (> 3.5 $\underline{M}$ HNO_3) nitric acid! The eluted plutonium solution
is sent elsewhere in the plutonium facility for further processing. The anion resin is frequently replaced with new resin to
prevent buildup of resin degradation products.

Before the americium can be precipitated as the oxalate,
the acidity of the solution must be lowered. This cannot be
done by the addition of NaOH or KOH as these cations are carried down with the americium oxalate. The acidity adjustment
can be made with NH_4OH with no product contamination, but processing problems resulting from ammonia vapors mixing with
nitric acid fumes have to be avoided. Even with the use of
efficient traps, some ammonia vapors escape to form solid ammonium nitrate which plugs glovebox exhaust filters; plus,
ammonium nitrate also slowly sublimes through the entire exhaust system.

The above mentioned problems make it highly desirable to
lower the acidity of the ion exchange effluent with only distilled water. Volume constraints of equipment make it imprac-
tical to dilute 7 $\underline{M}$ HNO_3 down to 0.5 $\underline{M}$ HNO_3 solution. Some
method of denitrification must be used to remove HNO_3 before
the final adjustment is done with water. This denitrification
is accomplished by simple distillation (Figure 7). Approximately a ten-fold reduction, both in volume and total moles of HNO_3,
is achieved by distilling down 5 liter batches of the ion exchange effluent. Before the reduced volume of americium nitrate
solution completely cools to ambient temperature and salts out,
enough distilled water is added to keep it in solution. During
this volume reduction, the americium has also been concentrated,
and appropriate shielding must be used.

The concentrated americium nitrate solution, now approxi-
mately 2 $\underline{M}$ in HNO_3, is transferred by vacuum into a stainless
steel 2 liter bottle and transported to the oxalate precipita-

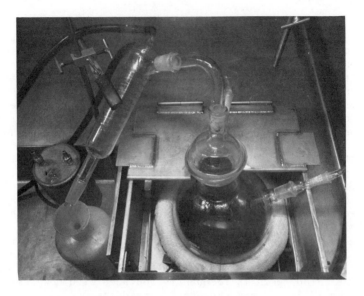

Figure 7. Denitrification apparatus (including shielding) for americium nitrate
solution

Figure 8.   Americium oxalate precipitation vessels

tion process. The americium oxalate precipitation vessel is shown in Figure 8. The americium nitrate solution is transferred into the precipitation vessel by vacuum. After titration, the final acidity adjustment to 0.5 $\underline{M}$ is made with the addition of distilled water. The americium is then precipitated with an excess of oxalic acid. At present, no in-line instrument is available for determining the exact concentration of americium in each batch; therefore, an excess of oxalic acid must be used to insure complete precipitation. These conditions allow some of the americium to be lost as a soluble oxalato complex whenever excess oxalic acid is available.

The americium oxalate is allowed to digest, with stirring, for at least one hour to allow time for post precipitation. After digestion, the americium oxalate is filtered onto a Kynar (Pennwalt Company) frit in a stainless steel filter boat (Figure 9). The precipitate is washed with 0.1 $\underline{M}$ oxalic acid and dried by pulling air through the frit.

Calcination of the americium oxalate to AmO_2 is done in two stages. The americium oxalate is first heated in a quartz beaker on a flask heater, gently at first to prevent material from being carried out of the beaker by the expulsion of decomposition products of oxalate and then more strongly until the yellow oxalate has entirely blackened. A final calcination of the mixture is done in a platinum dish inside a muffle furnace at 800°C for four hours to insure complete conversion to AmO_2. If necessary, the AmO_2 is quickly ground to a fine powder in a mortar and pestle and resubmitted for calcination in the muffle furnace.

Finally, the black-brown americium oxide is sieved through a 170 mesh 7.6 cm screen on a mechanical shaker. A complete analysis is done to determine whether the AmO_2 meets the required product specifications. The analysis of representative batches of americium oxide produced at LASL is shown in Table II. Radiochemistry, emission spectroscopy, calorimetry, and spark source mass spectrometry are used in the analysis of the AmO_2. Providing the AmO_2 meets all of the product specifications, it is prepared for shipment. Product not meeting these specifications is recycled through the process at the proper entry point.

Americium oxide prepared at Los Alamos is shipped in the configuration shown in Figure 10. The AmO_2, in 25 gram quantities, is placed inside the stainless steel container which is then decontaminated and placed into a plastic bag. This is bagged out of the glovebox, placed in an open hood, and decontaminated once again. After being placed in a cold plastic bag and taped, it is then wrapped in steel wool and canned in a lead lined food pack can. This can is canned in a second food pack can. This configuration is stored in the vault until shipment.

T A B L E   II   PRODUCT ANALYSIS OF LASL AmO_2

Results are in ppm except for Am and Pu which are reported in Wt. %.

Batch No. Element	AmO_2 - 20	AmO_2 - 24	AmO_2 - 27
Am	87.3 %	87.5 %	86.0 %
Pu	0.076%	2.2 %	0.32 %
Al	7	22	7
Ca	27	840	24
Cl	20	35	27
Cr	19	60	25
F	3	5	17
Fe	40	73	53
K	12	24	8
Mg	2	4	17
Mn	2	6	3
Na	50	140	580
Ni	9	13	14
Np	500	450	450
P	3	1	6
Pb	370	34	5
S	--	5	--
Si	94	980	110
Th	--	74	--
U	< 10	365	38
Y	110	28	360

Figure 9.   Americium oxalate in stainless steel filter boat

Future Americium Production.

New Americium Source Stream. At Los Alamos, the effluent
from the Pu ion exchange columns contain salts resulting from
up-stream processing, $\sim$ 7 $\underline{M}$ HNO_3, Am, and small amounts of Pu.
This stream feeds the evaporators that produce recycle HNO_3
and results in a solution that is concentrated in salts that
include the Am. As this solution is cooled, some of the salts
crystallize out and are readily separated from the supernatant
liquid. This supernate, $\sim$ 7 - 9 $\underline{N}$ in HNO_3, is made alkaline
by NaOH addition causing most of the metals to precipitate as
the hydroxides. Filtration then gives a filtrate that can be
discarded plus a hydroxide cake. Presently, the hydroxide cake
and the crystallized salts are put into 20-year retrievable
storage.

It has been determined that the bulk of any Am in the original evaporator feed ends up in the supernate, hence in the
hydroxide cake. This is considered as a potential Am source
for the AmO_2 production line.

The supernate and hydroxide cake contain primarily Na, Al,
Mg, Ca, Fe, with some Pb, U, and Pu plus the Am. Therefore,
recovery of the Am from either the supernate or redissolved
hydroxide cake involves its separation from a highly salted
solution of these cations.

Advantage may be taken of these conditions by use of solvent extraction techniques. It is known (25-30) that Am extraction with TBP (tributyl phosphate) or DBBP (dibutyl butyl phosphonate) is enhanced by high nitrate salt concentrations in the
aqueous phase, particularly at HNO_3 concentrations below $\sim$ 1.0 $\underline{N}$.

Solvent Extraction Experiments. Solvent extraction
studies were done on two feed samples representing dissolved
hydroxide cake (SSA) and evaporator supernate (SSB). SSA was
prepared by dissolution of hydroxide cake with slow addition of
concentrated HNO_3, adjustment of the final acidity to $\sim$ 0.5 $\underline{N}$
by addition of water and/or HNO_3, and clarification by filtration. To prepare SSB, some supernate liquid from the evaporator
was titrated, acidity adjusted to $\sim$ 0.5 $\underline{N}$ by NaOH and water addition, then clarified by filtration. No appreciable solids
were observed on the clarification filters for either solution.
Compositions of these feeds are listed in Table III.

After exploratory tests on a solution simulating SSA but
with Nd as an Am stand-in, the extraction system 30 vol % DBBP-
70 vol % Isopar H (Mobil Chemical Company) was chosen for tests
on the SSA and SSB feeds.

All experiments were done in separatory funnels by mixing
equal volumes of the organic and aqueous phases for $\sim$ 5 minutes,
allowing to settle $\sim$ 5 minutes, and then separating. Samples
of each of the two phases were analyzed for the constituents of

TABLE III
FEED SOLUTION COMPOSITIONS

SPECIES	(Dissolved Hydroxide Cake) SSA		(Evaporator Supernate) SSB	
	g/L	M	g/L	M
Am	0.46		0.34	
Pu	0.15		0.029	
U	3.9		0.81	
Pb	1.3		0.45	
Th	---		0.004	
Na		1.0		1.3
Al		0.4		0.67
Ca		0.3		0.23
Fe		0.21		0.11
Mg		0.15		0.07
Cr		0.06		0.02
Ni		0.06		0.01
K		---		0.05
H^+		0.44		0.5
NO_3^-		~ 4.5		~ 5.0

Figure 10.   Shipping configuration for americium oxide

interest. All reported results are those of experiments that
displayed acceptable mass balances which were used as a check
of experimental validity.

Results and Discussion.

Extraction and stripping experiments were designed to provide data for several areas of interest.

• Am distribution coefficients, D_{Am}'s, as a function of the
  equilibrium aqueous phase acidity of the feed were obtained.
  (D_{Am} is defined as the ratio of the Am concentration in the
  organic phase to that in the aqueous phase when the phases
  are in equilibrium.) These results are shown in Figure 11.
  The D_{Am}'s vary from 6 to 60 for acidities from 0.4 to 0.03
  N, respectively. The data points denoted X are for solution SSB, that is higher in NO_3 salt concentration than
  SSA, and the curve is struck through the SSA solution data
  denoted as O's.

• Am distribution coefficients as a function of the equilibrium aqueous phase acidity of non-salted strip solutions
  were studied. The results are given in Figure 12. Here,
  the D_{Am} varies from 0.25 to 0.01 for acidities from 0.4 to
  0.03 N. These low values from about 0.1 N downward are
  most valuable for stripping.

• Acid distribution between the equilibrated organic and feed
  phases is shown in Figure 13. These data are necessary for
  maintaining proper acid conditions during the extraction
  process. Previous workers (30) have shown that HNO_3 forms
  a 1:1 complex with DBBP, so it is evident that both HNO_3
  and Am transfer to the extractant. Although this reduction
  of feed acidity helps Am extraction by increasing the D_{Am},
  excessive feed acidity reduction may cause solids to form
  in the system by hydrolysis of the metal salts; therefore,
  it must be kept under control.

• The extractive effectiveness of DBBP-Isopar H was tested
  under several conditions: as received; pre-acidified only;
  and washed with Na_2CO_3 solution and then pre-acidified.
  The results for all three systems were the same within
  experimental error.

• Tests of the reuse of the solvent system after the Am was
  stripped showed no anomalies in the D_{Am}.

• The feed solutions for the extractions contained appreciable amounts of Pb, U, and Pu, so analyses were done to
  determine their course through the extraction-strip cycles.
  It was found that with either water or 0.01 N HNO_3 strips,
  the Am product solution did not have appreciable Pb contamination; however, the Pu and the U contamination was
  unacceptably high. Stripping with higher concentrations
  of HNO_3 would offer two advantages; first, the extraction

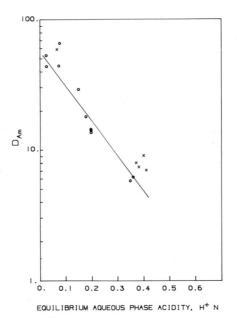

Figure 11. D_{Am} *vs. feed acidity for 30 vol % DBBP–70 vol % Isopar H at 21°C,
organic:aqueous = 1:1; (○) solution SSA, (×) solution SSB*

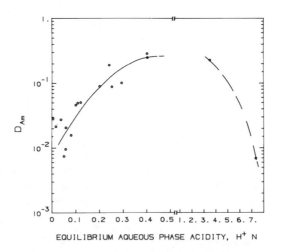

Figure 12. D_{Am} *vs. aqueous strip acidity for 30 vol % DBBP–70 vol % Isopar H
at 21°C, organic:aqueous = 1:1*

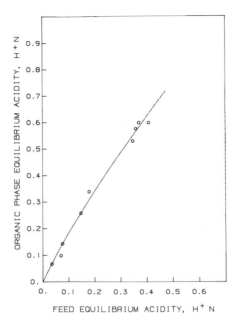

*Figure 13. Acid distribution between solvent and feed:solvent = 30 vol %
DBBP–70 vol % Isopar H, T = 21°C; organic:aqueous = 1:1*

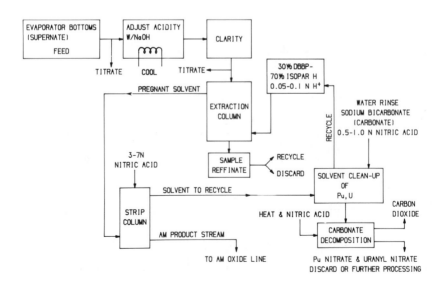

Figure 14. Proposed solvent extraction process for americium

coefficient for U and Pu would be larger, and second, the Am complexes with the HNO_3 in the aqueous phase at higher HNO_3 concentrations (26) therefore stripping better. Tests showed that contamination of the Am by Pu was essentially eliminated by stripping the Am with 3.65 to 7.3 $\underline{N}$ HNO_3 whereas that by U was reduced to acceptably low levels. The Am stripping D_{Am} for these strips was 0.22.

• Since the U and Pu are readily extracted from the aqueous feed and are forced to stay in the organic during the Am stripping, the solvent will eventually be loaded with them. Tests of solvent clean-up showed that these could be quickly and cleanly stripped with 0.5 $\underline{M}$ Na_2CO_3, 1.0 $\underline{M}$ $NaHCO_3$, or 2 $\underline{M}$ $(NH_4)_2CO_3$ solutions with no solid formation. The carbonate wash will also remove any degradation products from chemical and radiolytic attack on the DBBP.

Proposed Solvent Extraction Process for Am Recovery.

These studies resulted in the proposed solvent extraction recovery process for Am as given in the flowsheet of Figure 14. Solvent clean-up and recycle are included. The Am product stream is to enter the present AmO_2 production line either before or behind the ion exchange step (Figure 3) depending upon the concentration of Pu in the extracted americium.

Literature Cited.

1.   Stephanou, S. E.; Asprey, L. B.; and Penneman, R. A.; Report AECU-925, 1950.
2.   Penneman, R. A.; and Asprey, L. B., Report AECU-936, 1950.
3.   Asprey, L. B.; Penneman, R. A.; and Stephanou, S. E., Report AECU-927, 1950.
4.   Asprey, L. B.; Stephanou, S. E.; and Penneman, R. A., J. Amer. Chem. Soc., 1950, 72, 1425.
5.   Asprey, L. B.; Stephanou, S. E.; and Penneman, R. A., J. Amer. Chem. Soc., 1951, 73, 5715.
6.   Stephanou, S. E.; and Penneman, R. A., J. Amer. Chem. Soc., 1952, 74, 3701.
7.   Jones, Llewellyn H.; and Penneman, R. A., J. Chem. Phys., 1953, 21, 542.
8.   Keenan, T. K.; Penneman, R. A.; and McInteer, B. B., J. Chem. Phys., 1953, 21, 1802.
9.   Stephanou, S. E.; Nigon, J. P.; and Penneman, R. A., J. Chem. Phys., 1953, 21, 42.
10. Nigon, J. P.; Penneman, R. A.; Staritzky, E.; Keenan, T. K.; and Asprey, L. B., J. Phys. Chem., 1954, 58, 403.
11. Keenan, T. K.; Penneman, R. A.; and Suttle, John F.; J. Phys. Chem., 1955, 59, 381.

12. Armstrong, D. E.; Asprey, L. B.; Coleman, J. S.; Keenan, T. K.; LaMar, L. E.; and Penneman, R. A., Los Alamos Scientific Laboratory Report LA-1975, 1956.
13. Penneman, R. A.; and Asprey, L. B., Proc. Int'l. Conf. on Peaceful Uses of Atomic Energy, 1956, 7, 355.
14. Coleman, J. S.; Penneman, R. A., Keenan, T. K.; LaMar, L. E.; Armstrong, D. E.; and Asprey, L. B., J. Inorg. Nucl. Chem., 1957, 3, 327.
15. Armstrong, D. E.; Asprey, L. B.; Coleman, J. S.; Keenan, T. K.; LaMar, L. E.; and Penneman, R. A., AIChE Journal, 1957, 3, 286.
16. Penneman, R. A.; and Keenan, T. K., "The Radiochemistry of Americium and Curium," NAS-NS-3006, 1960.
17. Penneman, R. A.; Coleman, J. S.; and Keenan, T. K., J. Inorg. Nucl. Chem., 1961, 17, 138.
18. Asprey, L. B.; and Penneman, R. A., J. Amer. Chem. Soc., 1961, 83, 2200.
19. Asprey, L. B.; Penneman, R. A.; and Kruse, F. H., Chem. and Eng. News, 1962, 40, No. 8, 39.
20. Asprey, L. B.; and Penneman, R. A., Inorg. Chem., 1962, 1, 134.
21. Coleman, J. S.; Keenan, T. K.; Jones, L. H.; Carnall, W. T.; and Penneman, R. A., Inorg. Chem., 1963, 2, 58.
22. Penneman, R. A.; Kruse, F. H. Benz, R.; and Douglass, R. M., Inorg. Chem., 1963, 2, 799.
23. Asprey, L. B.; and Penneman, R. A., Chem. and Eng. News, 1967, 45, No. 32, 74.
24. Penneman, R. A.; Keenan, T. K.; and Asprey, L. B., "Lanthanide/Actinide Chemistry," Amer. Chem. Soc. Advances in Chem. Series, 1967, 71, 248.
25. Walsh, K. A., Los Alamos Scientific Laboratory Report, LA-1861, 1955.
26. Maraman, W. J., Los Alamos Scientific Laboratory Report, LA-1699, 1954.
27. Zemlyanukhin, U. I.; Savoskia, G. P.; and Pushlenkov, M. F., Sov. Radiochem. (Eng.), 1962, 4, 501.
28. Zemlyanukhin, U. I. Savoskia, G. P.; and Pushlenkov, M. F., Sov. Radiochem. (Eng.), 1964, 6, 673.
29. Sheppard, J. O., General Electric, Hanford Atomic Products Operations Report HW-81166, 1964.
30. Kinglsey, R. S., General Electric, Hanford Atomic Products Operation Report RL-SEP-518, 1965.

RECEIVED March 23, 1981.

Recovery of Americium-241 From Aged Plutonium Metal

L. W. GRAY, G. A. BURNEY, T.A. REILLY,
T. W. WILSON, and J. M. McKIBBEN

E. I. du Pont de Nemours & Company, Savannah River Laboratory, Aiken, SC 29808

The Savannah River Plant (SRP) was requested to separate approximately 5 kg of ^{241}Am from about 850 kg of plutonium metal containing, nominally, 11.5% ^{240}Pu. After separation and purification, both actinides were precipitated as oxalates and calcined to their respective oxides. The PuO_2 was shipped to the U.S. Department of Energy Hanford Site for use as fuel in the Fast Flux Test Facility (FFTF); the ^{241}AmO$_2$ was shipped to the Oak Ridge National Laboratory Isotope Sales Pool for use as a neutron source in many fields, predominantly petroleum well-logging.

A large-scale process was developed specifically for SRP application using established dissolution, separation, purification, precipitation, and calcination technology. However, adaptation of the process to existing plant facilities required a substantial development effort to control corrosion, to avoid product contamination, to keep the volume of process and waste solutions manageable, and to denitrate solutions with formic acid. The Multipurpose Processing Facility (MPPF), designed for recovery of transplutonium isotopes, was used for the first time for the precipitation and calcination of americium. Also, for the first time, large-scale formic acid denitration was performed in a canyon vessel at SRP.

Conceptual Process

Because it was necessary to use a process that would work in existing equipment, a process was designed (diagrammed in Figure 1) involving the following operations:

- **Dissolution.** Plutonium metal was dissolved in 1.67M sulfamic acid at about 25°C to 60 $\pm$10 g Pu/L. The PuO_2 coating on the surface of the metal plus the PuH$_x$ (where x = 2.0 to 2.7) produced from the reaction of $H_2(g)$ with plutonium metal formed a sludge which was collected and subsequently dissolved separately using hot 14M HNO_3 containing 0.2M KF.

0097-6156/81/0161-0093$05.00/0

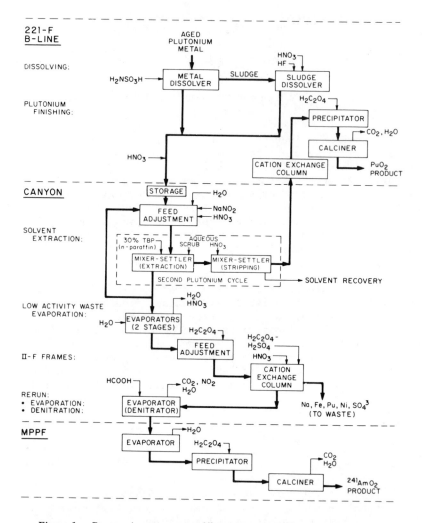

Figure 1.   Process for recovery of ^{241}Am from aged plutonium metal

● Feed Adjustment for Extraction. Dissolver solution was accumulated and diluted to ≤6 g Pu/L with 3M HNO₃. Sodium nitrite
was added to oxidize residual sulfamate and Pu(III) to Pu(IV).

● Solvent Extraction. Plutonium and americium were separated in
a single pass through a set of mixer-settlers using 30%
tri-n-butyl phosphate (TBP) in a n-paraffin hydrocarbon (NPH).

● Plutonium Finishing. The separated plutonium was processed to
PuO₂ by conventional cation resin exchange, oxalate precipitation, and calcination methods.

● Americium Feed Adjustment. The aqueous americium-bearing,
sodium nitrate/sulfate-nitric acid solution was evaporated and
acid-stripped. Oxalic acid was added to complex iron and
plutonium ions before feeding to a cation exchange column.

● Cation Exchange. Americium was further concentrated and purified by column chromatography using Dowex 50W x 8 resin.
After washing with H₂SO₄-H₂C₂O₄ solution, and then with
0.2M HNO₃, the americium was eluted with 5.5M HNO₃.

● Denitration. The eluted Am-HNO₃ solution was evaporated.
The nitric acid was reduced to about 0.25M by semibatch reaction with formic acid; the final concentration of americium
was about 2 g Am/L.

● Americium Finishing. The americium concentrate was precipitated in small batches by the addition of 0.9M oxalic acid,
digested at ambient temperature, filtered, washed, air-dried,
calcined at 700°C, and packaged for shipment.

Laboratory Demonstrations

After each processing unit operation, solutions actually
generated were used for laboratory demonstrations of the next
processing unit operation. Processing parameters were adjusted as
necessary to obtain a high yield of a high-purity product.

Pilot Demonstrations

After the laboratory demonstrations, plant scale demonstrations of formic acid denitration and of precipitation operations
were performed in both canyon and MPPF equipment with nonradioactive chemicals. These demonstration runs confirmed the
operating limits which had been established during the laboratory
experiments.

The cation exchange process, however, could not be piloted
with nonradioactive chemicals. Instead, each separate batch was

monitored by multiple sampling and rapid analysis of the raffinate
during the loading step. Subsequent column loadings from the same
batch of ^{241}Am solution were then adjusted according to the
losses determined after the first column-loading of each batch.

Process Yields

The process operated successfully in the plant. More than
98% of the americium was recovered from the cation exchange column
as an acidic nitrate solution. Substantial quantities of sodium,
iron, nickel, sulfate, and phosphate were removed. Decontamination from these impurities was satisfactory, but almost all the
chromium and small amounts of nickel, iron, and lead remained.
The plutonium metal feed stock contained about 5 ppm natural lead
which was not removed by the process. Recovery of americium in
the finishing process (oxalate precipitation and calcination)
averaged 98.5%. Most of the residual chromium contaminant was
removed from the oxalate in decanted supernate and washes. The
finished oxide product purity exceeded specifications; i.e., >95%
^{241}AmO$_2$. By selective blending, impurities in the shipped
product, predominantly lead and nickel, were kept below 2%.

Experimental Procedures

All experiments were conducted using normal laboratory glass-
ware. Chemicals used were technical grade chemicals removed
directly from process chemical hold tanks where possible; resins
were from the same production lots as would be placed in the
process column equipment. Cation exchange feed rates were the
same as obtainable in plant equipment.

Laboratory Results

Dissolution of Plutonium Metal

Plutonium metal dissolves readily in sulfamic acid (NH$_2$SO$_3$H)
at ambient temperatures according to the reaction

$$Pu^0 + 3H^+ \longrightarrow Pu^{+3} + 3/2\ H_2(g) \qquad (1)$$

The dissolution rate at about 25°C depends upon acid concentration
and surface area of the metal. Typically, initial batches of
solution from the dissolver average 50 $\pm$5 g Pu/L; the concentration increases to 60 $\pm$10 g Pu/L when using a cycle of 1 hour
dissolving time followed by displacement of two- thirds of the
solution. A more complete treatment of both ambient temperature
and elevated temperature dissolving experiments is given else-
where(1, 2, 3).

Storage of Dissolved Plutonium Solution

Simulated storage experiments showed (Figure 2) that radiolysis would be inadequate for valence adjustment of Pu(III) to
Pu(IV) within the available time frame. It was also necessary
to assure that plutonium sulfates would not precipitate during
storage. The solubility of plutonium vs. nitric acid concentration at various concentrations of sulfate is shown in Figure 3.
Because the plutonium concentration in canyon tanks is kept at
<6 g Pu/L, nitric acid concentrations as high as 6M can be toler-
ated as the sulfate ion concentration is diluted to <0.4M. while
diluting the Pu.

Feed Adjustment and Solvent Extraction

In the laboratory, the plutonium valence could be effectively adjusted to the extractable Pu(IV) state using N_2O_4,
NO, heat, or $NaNO_2$. For plant processing, $NaNO_2$ was chosen
as it is the method in routine use for normal SRP Purex processing. Heat was not used because sulfamate hydrolysis is more
rapid than its oxidation by nitric acid(4) and the hydrolysis
product, NH_4^+, has a higher affinity for a cation exchange
resin than does Na^+ ion. Gases were rejected because at present
a system is not available to add the gases to canyon tanks.
Chemically, either gas (N_2O_4 or NO) would have been more
desirable for downstream processing of the ^{241}Am.

Americium Concentration

The method of separation results in the ^{241}Am being diverted
to the aqueous waste stream of the second plutonium solvent extraction cycle (2 AW). The calculated predicted volume of this
stream for the full campaign was 2.2×10^6 L. Physical limitations of equipment required that the solution be evaporated in
two steps instead of one. The first step could result in a concentration factor of 25 to 50. The second step included nitric
acid stripping and evaporation to the final volume.

The solubilities of Na_2SO_4 and $(NH_4)_2SO_4$ as a function of
HNO_3 concentration (Figure 4) suggest hydrolysis of the sulfamate
ion to NH_4^+ and SO_4^{2-} would be better than oxidation with $NaNO_2$,
because the solubility of $NaHSO_4$ is exceeded during the first
stage of evaporation. However, downstream processing through
cation exchange dictates that oxidation, not hydrolysis, must be
the mode of destruction of the sulfamate ion.

Sodium americyl sulfate is also relatively insoluble in
nitric acid (Table I). The solubility of this salt is also exceeded during the first stage of evaporation. However, subsequent acid stripping of the solutions reduces the nitric acid
concentration and the salts redissolve.

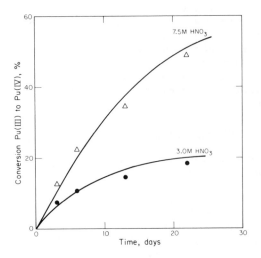

Figure 2. Radiolytic oxidation of Pu(III) to Pu(IV) in sulfamic acid–nitric acid solutions

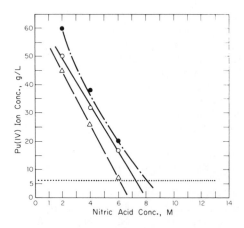

Figure 3. Calculated equilibrium concentrations of Pu(IV) in nitric acid solutions containing sulfate ions: (○) 0.25M H₂SO₄, (△ 0.40M H₂SO₄, (●) 0.2M H₂SO₄

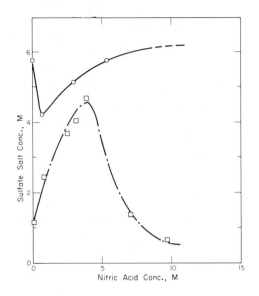

Figure 4. Solubility of sulfate salts in nitric acid solutions: (□) Na⁺ salts, (○) NH₄⁺ salts

TABLE I. Solubility of Sodium Americyl Sulfate

Composition of Solutions[a]			Am in solution, g/L (in equilibrium with solid)
HNO_3, M	Na^+, M	SO_4^{2-}, M	
0.8	2.1	1.0	0.15
1.4	2.3	1.1	0.25
1.4	3.3	1.5	0.15
4.2	2.1	1.0	0.4
4.2	3.1	1.5	0.2

a. All solutions contained Fe (1.5 g/L), Cr (0.4 g/L), and Ni (0.2 g/L).

The partial decomposition of dissolved TBP in the evaporation step leads to the precipitation of a white organophosphate
solid. Complete decomposition to an acid soluble compound required extended boiling of $\geq$10M nitric acid solutions as would be
achieved if the first evaporation step were a factor of 50. Use
of this evaporation factor in the first stage, however, led to
excessive corrosion of the stainless steel process equipment.

The first evaporation step was limited to a factor of 25 to
limit the HNO_3 concentration to $\leq$9M. The white solids product
plugged the ion exchange columns, but could be removed from the
column with a water wash.

Cation Exchange Experiments

Cation exchange experiments were performed with gel and
macroporous resins, and with both simulated and authentic plant
solutions. Scouting experiments showed that Dowex™ 50W x 8 was
the most effective cation resin tested. Complexing of Fe^{3+}
with oxalate ion was also necessary to obtain adequate column
capacity. Test data are summarized in Table II.

The resin capacity for americium was only a few percent of
theoretical due to large concentrations of other polyvalent
cations. The resin capacity was increased by adding one to two
moles of oxalic acid per mole of iron in the feed. Less than
10% of the iron was retained by the resin. Addition of oxalic
acid to both the feed and wash solution effectively separated
98% of the iron as well as >98% of the trace Zr, Nb, and Pu
ions.

Oxalic acid in the feed does not affect the sorption behavior of Cr^{3+} and Ni^{2+} ions on the cation resin. However,
about 75% of the Ni(II) ions were in the sorption and wash
effluents because resin affinity for Ni^{2+} ion is lower than
that for Cr^{3+} and Am^{3+} ions.

More than 99% of the sodium was separated when two acid
washes were made after the loading cycle. The first wash was
about 15 bed volumes of 0.2M H_2SO_4-0.05M $H_2C_2O_4$; the second
wash was about 5 bed volumes of 0.25M HNO_3, which removed the
remaining sodium and also flushed sulfate and oxalate from the
resin bed.

Elution with 5M HNO_3 at 0.5 mL/(min-cm^2) removed about
87% of the americium in four bed-volumes and >99% in eight
bed-volumes.

Formic Acid Denitration of Product Solution

Results of formic acid denitration simulation experiments (Figure 5) showed that at $\geq 90°C$, denitration began when sufficient 23.5M formic acid had been added to bring the solution concentration to about 0.06M formic acid. The lowest free-acid concentrations for the laboratory solutions were obtained when a formic acid-to-free nitric acid mole ratio of about 1.6 to 1.9 was used. This ratio yielded a final free-acid concentration of 0.7 to 0.8M. In the region where free nitric acid is $\leq 3M$, formic acid begins to accumulate in the solution. This accumulated formic acid is oxidized by refluxing the solution after all formic acid has been added. (Free acid is defined as the H^+ ion resulting only from strong acids, whereas total acid is the sum of free acid plus the H^+ ion resulting from the hydrolysis of hydrolyzable cations.)

To handle the volume of solution (about 30,000 L) necessary in the plant operation, a semi-batch denitration was necessary. Slow evaporation during product accumulation reduced the volume to <12,000 L, but increased the nitric acid concentration to about 11M. Experiments indicated that for a semi-batch denitration mode, a projected nitric acid concentration of 2M was an excellent stopping point, because no residual formic acid remains through the reflux and evaporation steps. Additional high nitric acid solution can then be added to the evaporated-denitrated solution without auto-initiation of a formic acid-nitric acid reaction. After all the Am-bearing solution had been transferred to the denitration evaporator and denitrated to $\leq 2M$, the solution could be evaporated to 2500 L and denitrated to a residual free-acid concentration of 0.5 to 0.8M. In actual practice, the final 2500 L of solution was denitrated to 0.25M HNO_3.

The lower acidity obtained in the plant-scale run resulted in the precipitation of a small amount of an iron organophosphate material which could be dissolved at 50°C in 0.5M HNO_3. Therefore, after moving the solution to the hold tank, the denitration evaporator was flushed with 1M HNO_3. This flush raised the acid concentration of the prepared solution to 0.37M and the volume to 2700 L.

Precipitation of ^{241}Am Oxalate

Americium is separated from iron, chromium, nickel, and other impurities by oxalate precipitation. The ^{241}Am feed solution for precipitation in the MPPF after formic acid denitration and volume reduction was approximately 2 g Am/L, 14 g Cr/L, 1.2 g Fe/L, and 0.8 g Ni/L in 1M HNO_3. Further concentration of the feed solution to >2 g Am/L necessitated evaporation at <85°C because of the potential corrosion of the stainless steel

TABLE II.   Capacity of Dowex   50W x 8 for Americium Retention

| | Feed Composition[a] | | | | Capacity of |
Test	Fe, g/L	Cr, g/L	Ni, g/L	Oxalic Acid, M	Resin, Bed Volumes
1	0.1	0.05	0.02	–	52
2	0.3	0.15	0.05	–	35
3	0.1	0.05	0.02	0.005	70
4	1.5	0.4	0.2	0.05	40
5	3.5	1.3	0.7	0.1	12
6	1.7	0.6	0.35	0.05	55

a.   Feeds for Tests 1 through 5 were $0.5M\ H^+$ – $0.4M\ Na^+$ – $0.25M\ SO_4^{2-}$ – 0.07 g Am/L. Feed for Test 6 was $0.25M\ H^{+4}$ – $0.2M\ Na^+$ – $0.125M\ SO_4^{2-}$ – 0.035 Am/L.

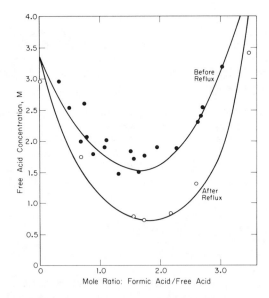

Figure 5.   Formic acid denitration of ^{241}Am solutions containing gross quantities of chromium nitrate

evaporator resulting from Cr(VI) in hot strong nitric acid. A precipitation test with plant solution adjusted to 2 g Am/L yielded americium oxide that met purity guidelines (Table III). Emission spectrograph analyses showed 0.25 wt % Pb, 0.15 wt % Ni, 0.14 wt % Cr, and 0.1 wt % Fe. All other impurities were <0.1 wt %, and the total impurities were 0.8 wt %. Another test with the plant solution yielded americium oxide with 1.6 wt % total impurities, and with nickel and chromium each about 0.5 wt %.

Table III. $^{241}AmO_2$ Purity Guidelines

Component	Final Product Composition, %
AmO_2[a]	>95
Pu	≤0.5
Pb	≤0.5
Any other element[b]	≤0.5

a. Analyzed by calorimetry.
b. Determined by emission spectrographic analyses.

Preliminary Plant Tests

Full-scale simulations of the formic acid denitration and the precipitation of rare earths (simulating americium) were carried out in plant equipment before processing the actual stream containing the ^{241}Am.

Denitration of Nitric Acid

Three simulated tests were run using only nitric and formic acids. In each case, the reaction began promptly and proceeded smoothly. After the second test, the denitrated material was evaporated and additional nitric acid added to simulate the tandem semi-batch operation to be used with actual process solution. At the end of formic acid feed for Test 3, the material was refluxed for 2 hours and evaporated to 2500 L to simulate the final canyon product batch. A final formic acid denitration reduced the acidity of the simulated concentrate to <1M HNO_3.

Oxalate Precipitation of Simulated Solutions

Two precipitation conditions were investigated with either dysprosium or samarium as a stand-in for ^{241}Am. Both conditions were based on physical limitations on the volumes in tanks to be used in the ^{241}Am solution processing. The denitrated product solution could be concentrated in the canyon evaporator; if so, the volume of solution required to just cover the steam coils in this evaporator, nearly 2500 L, would be the minimum volume. Hence, one condition assumed that the solution was evaporated to 2500 L, which would give 2 g Am/L. The second condition was set by the maximum volume of the evaporator; i.e., 1850 L. Hence, the second condition assumed the solution was further evaporated to 1800 L, which would give 2.8 g Am/L. Testing in the MPPF used concentrations of contaminant that would be found at both 1800 L and 2500 L. Testing in the MPPF confirmed that conditions assuming 2500 L would yield an acceptably pure product, whereas 1800 L might yield a marginally pure product.

Later laboratory demonstrations using actual solutions, however, showed that acceptably pure product could be precipitated from a 4 or 6 g Am/L solution, the equivalent of evaporating the solution to about 900 L. Emission spectrographic analyses showed the total impurities of the americium product from both the 4 and 6 g/L solutions to be about 1.5 wt %.

Plant Processing

Prior to beginning actual plant processing, it was necessary to flush all tanks and pipes extensively to avoid contamination of the plutonium and the americium with plutonium of a different isotopic composition or with fission products or other impurities. Processing began only after analyses of the flush solutions confirmed the product contamination would be acceptably low.

Plutonium Dissolving

Plutonium metal was dissolved in 1.67M sulfamic acid at an average rate of 1.81 kg per day per dissolver. Sludge and plutonium oxides generated from metal oxidation were dissolved in HNO_3-HF solutions. This plutonium concentrate (about 60 g Pu/L) was diluted to 5 to 6 g Pu/L before transfer to canyon storage to meet canyon nuclear safety requirements.

Solvent Extraction

Feed for the second plutonium cycle was prepared by first oxidizing the Pu(III) to Pu(IV) and the sulfamate ion to nitrogen gas and sulfate ion with sodium nitrite. The plutonium was diluted to about 0.5 g/L to meet the nuclear safety requirements of the second plutonium cycle. Nitric acid was adjusted to 3.8 to 4.0M to meet the salting requirements of the solvent extraction separation process.

Americium and plutonium were separated by one cycle of
solvent extraction using 30% TBP in a normal paraffin hydrocarbon
diluent. Plutonium was extracted while americium was diverted to
the aqueous waste stream (2AW).

Plutonium was stripped from the solvent with hydroxylamine,
concentrated further by cation exchange, precipitated as plutonium oxalate, and calcined to the oxide.

Evaporation and Steam Stripping

The aqueous waste stream (2AW) containing the ^{241}Am was
concentrated and stripped of acid using two batch evaporators in
the Low Activity Waste (LAW) system. The first concentration
step was performed in the LAW batch evaporator. Acid stripping
with water and additional evaporation was performed in the second
LAW batch evaporator. The average concentration of the ^{241}Am
entering this two-step evaporation process was 3.4×10^{-3} g/L;
after the first step, the ^{241}Am concentration was 0.08 to
0.15 g/L; after the second step, the ^{241}Am concentration was
0.2 to 0.3 g/L, and the nitric acid concentration was 2.0 to 2.5M.

Cation Exchange

Feed adjustment consisted of diluting the ^{241}Am concentrate
with water such that the concentration of hydrogen ion plus sodium
ion was less than or equal to one molar, and adding oxalic acid
(0.03-0.05M) to serve as a complexing agent to facilitate rejection
of Fe, ^{95}Zr, ^{95}Nb, and Pu ions.

The adjusted solution was fed to a 38.1 cm (15-in.)-diameter
cation exchange column filled with 42 L of Dowex 50W x 8 (50 to
100 mesh) resin. After the load cycle, the column was washed
first with 0.25M H_2SO_4 to further remove Na, ^{95}Zr, ^{95}Nb, Pu,
and Fe ions, and then with 0.25M HNO_3 to remove sulfate. Americium was eluted with 5M HNO_3 and the resin reconditioned with
dilute HNO_3 for the next run. A summary of typical column
operations is given in Table IV.

Slow evaporation (70 to 80°C) allowed storage of all the
product in one tank. This slow evaporation was successful in
controlling the corrosion due to chromium in the product.

The isolation system (evaporation plus cation exchange)
recovered >98% of the ^{241}Am, with a concentration factor of
125, while rejecting >96% of the Na$^+$, SO_4^{2-}, Fe^{3+}, Pu, and
fission product ions. However, the process rejected only 88.8%
of the nickel and <1% of the chromium.

Table IV. Typical Column Runs

Component	Feed	Product	Recovery, %
Am, g	40 - 60	39.2 - 58.8	>98
Pu, g	0.3 - 0.6	<0.01	<2
Na, g	7,000 - 14,000	25	0.3
$SO_4^=$, g	14,000 - 28,000	300	1
Fe, g	1,000 - 3,000	300 - 900	>99
Ni, g	150 - 450	15 - 100	20
Vol, L	1,500	325	-
H^+, M	0.4 - 0.6	5 - 6	-

Formic Acid Denitration

After the evaporation, approximately 36% of the [241]Am
solution was moved to a denitration evaporator. After dilution
from 11M to 8M HNO_3, the acidity of the solution was reduced
by reaction with formic acid to an estimated 3M HNO_3. After
refluxing to assure total destruction of the formic acid, the
denitrated solution was concentrated in the evaporator to about
55% of its original volume. A second transfer of [241]Am
solution from storage to the denitrated solution was made.

The denitration, refluxing and evaporation were repeated.
Four additional transfers, denitrations, refluxings, and evapora-
tions were necessary to move and concentrate all the solution to
about 2500 L. Then the final denitration of the entire batch was
carried out. All denitrations proceeded smoothly to completion.

Analysis of the final solution indicated no appreciable corrosion of the evaporator during the denitration procedure. The
final acidity was lower than that obtained in the laboratory
scale experiments, 0.25M versus 0.7M, respectively.

The lower acidity obtained in the plant scale run resulted
in the precipitation of a small amount of the iron, probably as
the phosphate. The precipitate was shown to dissolve, in the
laboratory, in 0.5M HNO_3 at 50°C. Therefore, after moving the
solution from the evaporator, the evaporator was flushed with 1M
HNO_3. This flush raised the acid concentration of the prepared
solution to about 0.37M. These flushes raised the volume to
2700 L.

Precipitation as Oxalate

Approximately 1500 L of solution was then transferred into the smaller MPPF evaporator. About 150 L of feed were further transferred into the MPPF as feed for the first four precipitator batches. The remaining 1350 L were simmered to reduce the volume to 500 L. Additional transfers were made to combine all the feed as well as flush the canyon tanks of all ^{241}Am products; simmering at 85°C continued so that all of the americium-containing solution would fit into the evaporator.

Precipitations were made by adding sufficient 0.9M oxalic acid to bring the final oxalate concentration to 0.3M. After a digestion period and decanting the filtrate, the oxalate precipitate was washed four times with 0.2M $H_2C_2O_4$-0.7M HNO_3, and once with 0.2M $(NH_4)_2C_2O_4$. On the initial runs, the washed oxalate precipitate was calcined to americium carbonate to allow easier acid dissolution if impurity analysis indicated recycle was required. As all product batches exceeded the purity guidelines, the low-temperature calcination step was eliminated and all products were calcined at 700°C.

Results for typical runs, all at about 2 g $^{241}Am/L$, are summarized in Table V. $^{241}AmO_2$ purity was very good (approximately 98% versus 95% minimum to meet guideline), and all impurities were insignificant except lead, which averaged 0.44% (guideline <0.5%), and weight loss, which averaged 0.59%.

Laboratory tests showed that the high weight loss was due to the sorption of water from the air during handling of the calcined powder.

TABLE V. Summary of ^{241}Am Recovery

Run Number	1	2	3	4	5
Product					
Gross product, g	56.6	64.7	52.2	72.3	92.7
^{241}Am, g	49.4	56.1	45.6	63.2	80.2
^{241}Am, %	87.3	86.7	87.3	87.0	86.5
$^{241}AmO_2$, g	56.0	63.6	51.6	71.6	90.8
$^{241}AmO_2$, %	98.8	98.2	98.9	98.5	98.0
Impurities					
Cr, %	0.02	0.02	0.01	0.02	0.04
Fe, %	0.05	0.06	0.07	0.03	0.05
Ni, %	0.04	0.10	0.10	0.02	0.01
Pb, %	0.40	0.40	0.50	0.40	0.35
C, ppm	<100	344	313	302	110
Weight loss, %	0.39	0.63	0.72	0.70	0.54

Acknowledgement

The information contained in this article was developed during the course of work under Contract No. DE-AC09-76-SR00001 with the U.S. Department of Energy.

Literature Cited

1. Gray, L. W., "The Kinetics of the Ambient Temperature Disso-
    lution of Plutonium Metal in Sulfamic Acid," USDOE Report
    DP-1484, E. I. du Pont de Nemours & Co., Savannah River
    Laboratory, Aiken, S.C., 1978.

2. Gray, L. W., "Dissolution of Plutonium Metal in Sulfamic Acid
    at Elevated Temperatures," USDOE Report DP-1515, E. I.
    du Pont de Nemours & Co., Savannah River Laboratory, Aiken,
    S.C., 1979.

3. Gray, L. W., "Rapid Dissolution of Plutonium Metal in
    Sulfamic Acid Followed by Conversion to a Nitric Acid
    Medium," accepted for publication in Nucl. Technol., MS.
    3826.

4. Gray, L. W., "The Interactions of Hydrazine, Ferrous Sulfa-
    mate, Sodium Nitrite, and Nitric Acid in Nuclear Fuel
    Processing Solutions," Nucl. Technol., 1978, 40 (185-193).

RECEIVED January 5, 1981.

Solvent Extraction Process for Recovery of Americium-241 at Hanford

P. C. DOTO, L. E. BRUNS, and W. W. SCHULZ

Rockwell Hanford Operations, Richland, WA 99352

Solvent extraction [tributyl phosphate (TBP)] operations to recover plutonium from unirradiated scrap have been performed at the Hanford Site since 1955. The aqueous raffinate (CAW stream) from the TBP plutonium extraction process contains, typically, <10 mg/L plutonium and 2-10 mg/L ^{241}Am. This latter isotope is present in the plutonium scrap as the result of beta decay of ^{241}Pu ($t_{1/2}$ = 14.4 y); trivalent americium does not accompany plutonium from the HNO$_3$-HF feed solution into the TBP solvent. In the early years of plutonium scrap processing operations, the CAW stream was routed to trenches(1) specially excavated in Hanford soil. Batch recovery of americium was started in 1965. Later (1970-1976), a continuous countercurrent solvent extraction process employing DBBP (dibutylbutyl phosphonate) as the extractant was operated to recover, at least partially, plutonium and americium values from the CAW stream. Aqueous waste from the DBBP extraction process, still containing some plutonium and americium, was blended with other Plutonium Reclamation Facility (PRF) wastes, made alkaline, and routed to underground tanks for storage.

In this paper we describe the Hanford DBBP ^{241}Am extraction process highlighting process chemistry, process equipment and facilities, and operating experience. Related research studies concerning laboratory tests of other extractants and solvent extraction processes for recovery of ^{241}Am at Hanford are also discussed. Discussion of the Hanford DBBP americium extraction process at this Symposium is particularly appropriate since it is, to date, the only countercurrent solvent extraction process operated routinely on a plant-scale for ^{241}Am recovery. (Other papers at this Symposium describe use of ion exchange processes for recovery of ^{241}Am.) The Hanford DBBP americium extraction process was last operated in 1976 and is not expected to be operated again since the PRF itself is scheduled to be shut down in the early 1980's.

0097-6156/81/0161-0109$05.25/0

Plutonium Reclamation Facility

In April of 1955, a facility utilizing the Recuplex solvent extraction process(2) was installed in the 234-5Z Building at Hanford. This facility provided the capability of recovering plutonium from unirradiated plutonium scrap from Hanford plutonium processing operations. By 1960, the Recuplex facility was inadequate with respect to contemplated production loads, shielding requirements, and criticality prevention safety. A project was authorized in March 1961 to provide a new facility for the adequate reclamation of plutonium from both wet and dry plutonium scrap generated from both on and offsite operations. This facility, the 236-Z Building, was completed in June 1964 and is referred to as the PRF. Details on the new plant were first published in 1967.(3)

The PRF solvent extraction process flowsheet is shown in Figure 1.(3,4) Initially, plant feed consisted solely of unirradiated plutonium scrap generated by other Hanford Site processing operations. Later, the plant was also used to recover plutonium from unirradiated scrap generated offsite. Various feed preparation processes including dissolution (in HNO₃-HF media) and/or leaching of solid scrap materials and concentration of liquids are utilized to prepare aqueous feed solutions. After addition of Al(NO₃)₃ to provide salting strength, Pu(IV) is extracted in the CA Column with a 20 vol% TBP-CCl₄ solvent.

Plutonium is subsequently stripped to an aqueous phase containing NH₂OH·HNO₃ in the CC Column. In order to increase the plutonium concentration of the CC Column product, a portion of this stream (CAIS) is recycled to the CA Column after adjustment with HNO₃. The remainder of the stream (CCP) is routed to the product concentrator. The resulting concentrated and purified plutonium nitrate solution is suitable feed to other processes for conversion to the desired product form (e.g., metal or plutonium dioxide). The remainder of the PRF solvent extraction system consists of a series of columns to wash the TBP-CCl₄ solvent and prepare it for reuse.

The aqueous waste from the CA Column (CAW) contains virtually all of the americium present in the feed. Table I shows the typical composition of the CAW stream. The exact composition of the stream depends on the composition of the CAF which is highly variable in a plutonium scrap processing plant. The CAW stream is the feed to the americium recovery solvent extraction system.

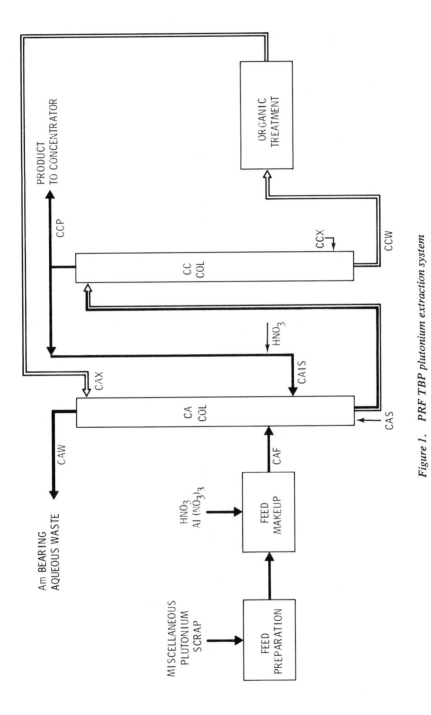

Figure 1. PRF TBP plutonium extraction system

TABLE I. Typical Composition
of Hanford CAW Solution.

Component	Concentration
NO_3^-	5.0$\underline{M}$
H^+	2.2$\underline{M}$
Al^{3+}	0.8$\underline{M}$
Na^+	0.5$\underline{M}$
F^-	0.3$\underline{M}$
Fe^{3+}	0.009$\underline{M}$
Si	0.002$\underline{M}$
Ca^{2+}	0.001$\underline{M}$
Cr^{3+}	0.0007$\underline{M}$
Mg^{2+}	0.0006$\underline{M}$
Ni^{2+}	0.0003$\underline{M}$
Pu	2-10 mg/L
^{241}Am	2-10 mg/L

Americium Extraction System

Chemistry. Dibutylbutyl phosphonate [DBBP =
$(C_4H_9O)_2(C_4H_9)PO$]- diluent solutions extract Am(III) and HNO_3
from aqueous nitrate media according to the reactions shown in
Equations (1) and (2), respectively:

$$Am^{3+}_{(aq)} + 3NO_3^-{}_{(aq)} + 3DBBP_{(org)} \overset{\rightarrow}{\leftarrow} Am(NO_3)_3 \cdot 3DBBP_{(org)} \qquad (1)$$

$$H^+_{(aq)} + NO_3^-{}_{(aq)} + DBBP_{(org)} \overset{\rightarrow}{\leftarrow} DBBP \cdot HNO_3{}_{(org)} \qquad (2)$$

The stoichiometry shown in Equation (1) is similar to that followed during extraction of trivalent americium by TBP and other
monofunctional neutral organophosphorus extractants. Distribution ratio data plotted in Figure 2 show that DBBP extracts
Am(III) more strongly than TBP from HNO_3 media. The equilibrium
constant for Reaction (1) (at zero ionic strength) is 7.4
compared to a value of only 0.4 for the similar reaction with
TBP.([5])
        From Equations (1) and (2), it is clear that in extraction
of americium from strong HNO_3 solutions, Am(III) must compete
with HNO_3 for available DBBP molecules. Sheppard's data,([6])
reproduced in Table II, are in line with this observation and

TABLE II. The Distribution of ^{241}Am Between 10% DBBP in Xylene and Nitric Acid - Sodium Nitrate - Metal Nitrate Solutions.*

$[HNO_3]$**	$[NaNO_3]$**	$[Al^{3+}]$**	$[Mg^{2+}]$**	$[Ca^{2+}]$**	$[Fe^{3+}]$**	D_{Am}
0.18	3.82	0.33				1.00
0.18	2.82	0.67				1.42
0.18	2.69	0.75				1.70
0.18	1.82	1.00				2.3
0.18	1.22	1.20				3.1
0.18	0.82	1.33				3.80
0.18	0.32	1.50				4.4
0.18		1.61				5.8
0.18		1.72				10.4
0.18		1.94				18.5
0.18	3.22		0.80			0.90
0.18	1.62		1.60			1.50
0.18	4.22			0.30		0.60
0.18	3.22			0.80		0.35
0.18	1.62			1.60		0.57
0.18				2.40		0.83
0.18					1.60	0.29

*Data of J. C. Sheppard, Reference 6.
**Molar concentration.

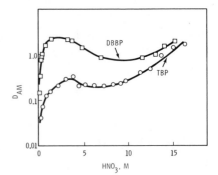

Figure 2.   Extraction of Am(III) by
TBP and DBBP from HNO₃

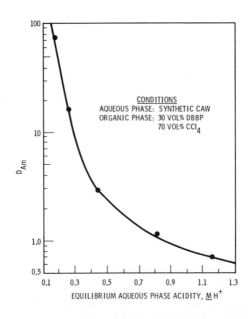

Figure 3.   Effect of CAW equilibrium aqueous phase acidity on DBBP extraction
of Am(III)

clearly demonstrate that distribution of Am(III) between DBBP solutions and aqueous nitrate solutions is greatly increased by the combination of high metal nitrate ion concentration and low nitric acid concentration. According to Sheppard, Equation (3) formally describes the distribution of ^{241}Am between HNO_3-metal nitrate solutions and DBBP-xylene solutions:

$$D_{Am} = [DBBP]^3 \left\{ 5.5 \times 10^{-6} \frac{[NO_3^-]^2}{[H^+]} + 1.7 \times 10^{-3}[Al^{3+}]^2 + 3.0 \times 10^{-4}[Mg^{2+}]^2 \right\} \tag{3}$$

In Equation (3) the DBBP concentration is expressed in vol%; all other concentrations are molarities. The nitrate ion concentration is the total from all sources. According to Sheppard, Equation (3) holds fairly well for 5M total nitrate ion, but should be considered only approximate at 4 and 6M total nitrate ion.

At constant nitrate ion concentration, Sheppard using a 10 vol% DBBP-xylene solvent found that the order of effectiveness of metal nitrates as salting out agents for extraction of Am^{3+} was $Al^{3+} > Mg^{+2} > Ca^{2+} \stackrel{\sim}{=} Na^+ > Fe^{3+}$. Kingsley[7], however, reported that for the system 30 vol% DBBP-CCl$_4$-synthetic CAW solution the order of effectiveness of various metal nitrates as salting agents was $Li^+ > Mg^{2+} > Al^{3+} > Na^+ > Ca^{2+}$. In any event, $Al(NO_3)_3$ added originally to the feed to the PRF TBP extraction process provides most of the salting strength required for DBBP extraction of Am^{3+} from neutralized CAW solution. Additional salting strength comes from $Ca(NO_3)_2$, $Mg(NO_3)_2$, and $Fe(NO_3)_3$ present in the CAW solution and from the $NaNO_3$ resulting when the CAW stream is adjusted to 0.1 - 0.3M HNO_3 by addition of NaOH.

Proper adjustment (neutralization) of CAW solution acidity is crucial to satisfactory DBBP extraction of trivalent americium. Kingsley's[7] results (Figure 3) show that the CAW acidity must be adjusted to a concentration in the range 0.1-0.3M to achieve adequate distribution of Am^{3+} to the DBBP phase. Proper adjustment of CAW acidity within this narrow range obviously requires careful control of in-line addition of concentrated NaOH to an unbuffered solution. Over-addition of NaOH precipitates hydroxide solids which scavenge actinides and interfere greatly with phase dispersion and separation. Insufficient addition of NaOH leaves, of course, excess HNO_3 to compete with americium for DBBP extractant.

Siddall[8] and others have observed that DBBP-diluent solutions extract Pu(IV) exceedingly well from both strong HNO_3 solutions and from metal nitrate-low HNO_3 solutions. The greater affinity of DBBP for Pu(IV) over that for Am(III) makes it possible to use dilute (0.1-0.25M) HNO_3 solutions to selectively partition americium from co-extracted plutonium. The resulting

acid strip solution is well conditioned as feed to subsequent
cation exchange resin concentration and purification of the
^{241}Am. Plutonium(IV) and residual ^{241}Am can be effectively
stripped from the DBBP phase by contact with a dilute HNO$_3$-HF
solution. Spent HNO$_3$-HF strip solutions can be conveniently
recycled to the feed preparation step of the mainline PRF TBP
extraction process.

The DBBP extraction scheme provides excellent decontamination of plutonium and americium from all the other metals in the
neutralized CAW solution. Kingsley(7) reports that distribution
ratios for Fe^{3+}, Al^{3+}, Ca^{2+}, and Mg^{2+} between 30 vol% DBBP-CCl$_4$
and neutralized CAW are, respectively, 0.11, <0.003, 0.025, and
<0.0005. Primarily because of entrainment but partly because of
extraction, small amounts of aluminum, iron, and sodium accompany
^{241}Am into the dilute HNO$_3$ strip solution. Richardson(9) in
nonradioactive tests of the countercurrent DBBP ^{241}Am recovery
process observed decontamination factors in the range 80-180
for iron and in the range 2 x 10^3 - 1.5 x 10^5 for aluminum.

Process Flowsheet. Figure 4 illustrates the typical chemical flowsheet conditions employed at Hanford in countercurrent
DBBP extraction of ^{241}Am from CAW solution. Over the six years
the process operated, there were, naturally, minor changes in
stream compositions and flowrates, but the values cited in Figure 4 are representative of those generally used. Chemical
flowsheets employed in plant-scale operations were based on the
previous results of Richardson(9) and Taylor(10).

Significant features of the flowsheet shown in Figure 4
include:

● Two-stage adjustment of CAW acidity to a value in the
   range 0.1 to 0.3M. Equipment and procedures used in
   this two-stage acidity adjustment step are described
   later.

● Countercurrent contact of the neutralized CAW solution
   (E1F stream) with 30% DBBP-CCl$_4$ to extract 99% of the
   soluble plutonium but only 66 to 80% of the ^{241}Am.
   Reasons, chiefly equipment limitations, which prevented
   extraction recovery of the desired 95-100% of the
   americium are outlined later.

● Contact of the DBBP extract in the WS-1 Column with a
   small flow of 0.1M HNO$_3$ to strip essentially all the
   extracted americium as well as about 30% of the
   plutonium.

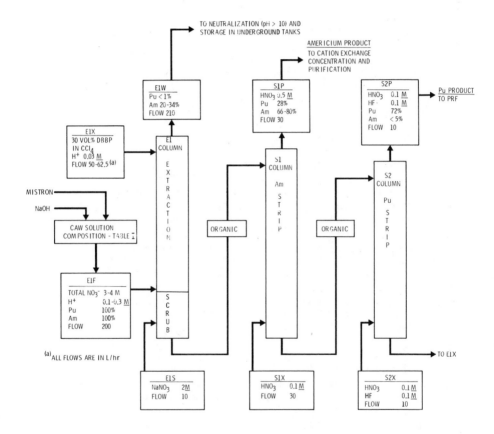

Figure 4. DBBP flowsheet for recovery of ^{241}Am from CAW solution

● Stripping of the residual plutonium in the WS-2 Column
  with a small volume of 0.1M HNO₃-0.1M HF solution. The
  resulting plutonium product is suitable for direct
  recycle to the mainline PRF TBP extraction process
  (Figure 1).

   Mistron is a tradename (Sierra Talc Co.) for finely divided
$MgSiO_3$ (Talc). Small amounts (e.g., 100-200 mg/L) of Mistron are
routinely added to the CA Column in the PRF TBP extraction process to aid in coalescence of emulsified organic and aqueous
phases. On occasions, Mistron was added to the WE-1 Column in
the DBBP extraction system to promote phase disengaging.
   A small flow of 2M NaNO₃ was routinely introduced to the
bottom of the WE-1 Column to scrub aluminum, iron, and other
metallic contaminants from the DBBP phase. Even so, americium in
the S1P stream is extensively contaminated with plutonium, aluminum, iron and other metallic impurities. Ion exchange procedures, both simple load-elute and chromatographic, are used
to yield highly purified ^{241}Am.
   Following an engineering analysis of the performance of the
DBBP extraction process up to mid-1973, Hammelman(11) proposed
alternative flowsheet conditions to improve extraction of ^{241}Am
and to improve separation of ^{241}Am from coextracted plutonium.
Major flowsheet changes proposed by Hammelman included substantially increased E1X stream flow (125 vice 50 L/hr), increased
S1X acidity (0.3M vice 0.1M HNO₃), and introduction of an
S1S Stream (30% DDBP-CCl₄ at 25 L/hr) to the top of the WS-1
Column. Hammelman calculated these alternative flowsheet conditions would provide for extraction of 98% of the americium in the
WE-1 Column and for production in the WS-1 Column of an americium
product containing only 1.2% of the plutonium. Equipment limita-
tions prevented implementation of the changes recommended by
Hammelman.

   Facility Description. Extraction columns used in both the
DBBP americium extraction process and in the mainline TBP plutonium extraction system are located in the same canyon facility
(Figure 5) in the 236-Z Building at the Hanford Site. Equipment
used to adjust the acidity of the CAW stream for subsequent
americium recovery is also located in the 236-Z Building. Sup-
porting tankage for the DBBP solvent extraction columns and ion
exchange columns for concentration and partial purification of
the recovered ^{241}Am are located in an adjacent building.
   Adjustment of the acidity of the CAW solution is performed
in two stages. The first stage, preneutralization, is carried
out in an 18-L glass tank (Figure 6A). Centrifuged CAW flows
into Tank 32-A from the CA Column where it is constantly recirculated to mix with added concentrated NaOH solution. Addition
of NaOH is controlled by electrical conductivity instrumentation
on the recirculation leg. Preneutralized solution (∿0.5M H⁺)

Figure 5. View of PRF canyon area

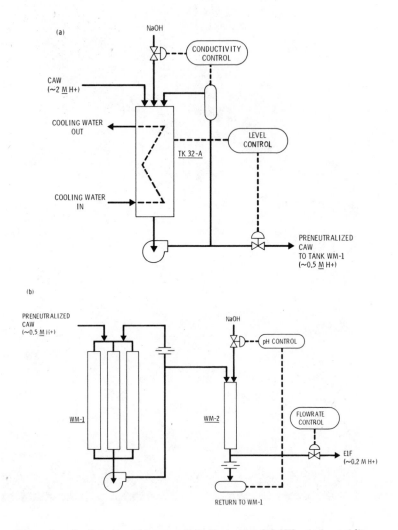

Figure 6.   Equipment and systems for adjustment of CAW solution acidity

flows from Tank 32-A via a level-controlled outlet valve to Tank WM-1, a 230-L three-barrel, geometrically favorable tank (for criticality control) which feeds the DBBP solvent extraction system.

The final pH adjustment (Figure 6B) takes place in the feed system to the extraction (WE-1) column. Preneutralized solution in Tank WM-1 is recirculated to provide mixing. Part of the recirculated solution is diverted to the WM-2 static mixer tank where it is mixed with NaOH; the volume of NaOH added is controlled by the pH of the WM-2 tank outlet stream. Most of the solution leaving the WM-2 tank is returned to the WM-1 tank; the remaining portion feeds the WE-1 Column. The flowrate of this latter stream is adjusted to maintain the liquid level in the WM-1 tank approximately constant. When functioning satisfactorily, the two-stage acid adjustment procedure provides an aqueous E1F stream at pH 0.75 flowing at a constant rate to the extraction column.

Figure 7 illustrates in schematic fashion the three column DBBP solvent extraction system. All three columns are made of 304L stainless steel columns and are packed with 1-inch Kynar (Pennwalt Corp.) Intalox saddles. Aqueous and organic phases are mixed by means of air pulses supplied to each column. Important column dimensions are shown in Figure 7.

Operation and Experience. A batch DBBP americium extraction process was operated for a short time prior to startup of the continuous countercurrent [241]Am extraction system. The batch process utilized much of the same chemistry as later used in the countercurrent system. Thus, large batches of CAW solution were adjusted to pH 0.5 to 1.0 and contacted with one-half volume portions of 30 vol% DBBP-CCl$_4$ solvent. Approximately 50% of the americium in the CAW reported to the DBBP phase. Essentially all the extracted [241]Am was stripped into an equal volume of 0.1M HNO$_3$ solution. Americium in the strip solution was sorbed onto a bed of Dowex 50W cation exchange resin and eluted with 6M HNO$_3$. Overall, about 40% of the [241]Am in the CAW feed was recovered; the recovered americium was heavily contaminated with aluminum, calcium, magnesium, and plutonium.

The countercurrent DBBP [241]Am extraction process was operated on a plant-scale for about six years. During that time, it provided excellent recovery (∼100%) of soluble plutonium in the feed and adequate decontamination of [241]Am from plutonium and other metallic impurities. The Am/Pu ratio in the WS-1 Column (Am strip column) product was 2.5/1 compared to only 1/1 for the batch extraction process americium product. Concentrations of calcium, aluminum, and other metallic impurities in the countercurrent Am product stream were also much lower than in the Am product from the batch extraction process.

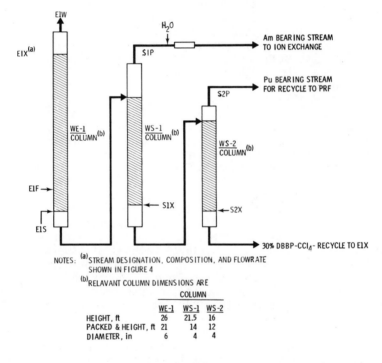

NOTES: (a)STREAM DESIGNATION, COMPOSITION, AND FLOWRATE
SHOWN IN FIGURE 4

(b)RELAVANT COLUMN DIMENSIONS ARE

	COLUMN		
	WE-1	WS-1	WS-2
HEIGHT, ft	26	21.5	16
PACKED & HEIGHT, ft	21	14	12
DIAMETER, in	6	4	4

Figure 7.   Americium solvent extraction columns

Over its six years of operation, the countercurrent DBBP extraction facility and process recovered from 66% to a high of 80% of the ^{241}Am in the neutralized CAW. These recoveries, although substantially higher than those realized in the batch process, fell short of those expected from the pilot plant tests with synthetic feeds. Engineering analyses conducted during process operation and after process shutdown indicate three primary factors contributed to the disappointingly low americium recoveries:

- Inadequate extraction equipment

- Inadequate feed pH adjustment

- Solids and organic impurities in actual CAW solution.

It was necessary, because of economic restraints, to locate the americium extraction battery in the same canyon as the main-line PRF TBP extraction equipment. Compromises in americium extraction equipment design and hydraulics were mandated to accommodate the americium extraction system in this existing facility. In particular, all the columns had to be shortened from the optimum heights recommended by the pilot plant studies. The hydraulics of the installed system were such that the organic flowrate to the extraction column (WE-1 Column) was inadequate with the result that the extraction factor was too low to permit quantitative extraction of all the ^{241}Am. Furthermore, the combination of a short extraction column and a 1-inch air pulse leg sometimes led to a hydraulic upset whereby the entire column contents were lost; recovery from such hydraulic upsets required an hour or more.

Three other deficiencies in the installed DBBP ^{241}Am extraction equipment were pointed out by Hammelman.(11) The ratio of packing diameter to column diameter is 1:6 for the WE-1 Column and 1:4 for both the WS-1 and WS-2 Columns. Both ratios are higher than the recommended maximum value of 1:8; incorrect balancing of packing and column diameters results in column channeling and contributes to reduced column efficiency. Secondly, redistributors and, to some extent, the initial distributors do not introduce the dispersed phase directly into the packing. The dispersed phase thus tends to coalesce when it contacts the packing support and this coalescence, greater because of reduced interfacial area for mass transfer, results in greater transfer unit heights. Finally, since the Kynar saddles are wetted by the organic phase, all three columns should be operated with the organic phase as the continuous phase. In actual practice the WS-1 and WS-2 Columns are operated with bottom interfaces and with the aqueous phase as the continuous phase. This latter type of operation leads to phase coalescence, filming of the organic phase, and greater transfer unit heights.

Considerable difficulty was experienced throughout the entire period of plant-scale operation of the DBBP countercurrent extraction process in adjusting the CAW solution to the desired pH of 0.75. Several factors contributed to these difficulties. Lack of any buffering capacity in the CAW solution made it easy to overshoot or undershoot the desired pH. The two-step neutralization procedure and equipment aided considerably in achieving proper feed acidity. But, even with this approach, inadequate mixing coupled with unsophisticated and insensitive monitoring and control instrumentation made it impossible to routinely achieve reliable adjustment of feed acidity to its optimum range.

Various finely divided solids including some PuO_2 are typically present in CAW solution. Such solids represent feed scrap materials not completely soluble in the HNO_3-HF dissolvent or which slowly precipitate from the feed to the PRF TBP process. Small amounts of organic materials (e.g., TBP degradation products) are also present in the CAW solution. Solid and organic impurities in the CAW solution contributed directly to flooding problems observed in the americium extraction battery. Attempts to control flooding in the WE-1 Column by regulation of the pulse amplitude while still maintaining flowrates set by the TBP extraction system were only partially successful. Oftentimes, to avoid flooding, the pulse amplitude was adjusted so low that phases were poorly mixed with concomitant increases in transfer unit heights and decreases in [241]Am extraction efficiency.

Co-location of the TBP and DBBP extraction processes in the same facility led inevitably to cross contamination of extractants. This problem was of greater consequence to the PRF system where small concentrations of DBBP in the TBP extractant interfered with plutonium stripping. No specific system malfunctions directly attributable to the presence of TBP in the DBBP solvent were identified. However, dilution of the DBBP extractant with TBP reduces its efficiency as an americium extractant.

Ion Exchange Purification of Recovered [241]Am. A simple load-elute cation exchange resin step was used to concentrate and partially purify [241]Am recovered by the DBBP extraction scheme. For this step the S1P Stream (Figure 4) was diluted with water to about $0.24\underline{M}$ HNO_3 and then loaded at 25°C onto a 14-liter bed (15-cm diameter) of H^+-form Dowex 50-X8 cation exchange resin. Considerable decontamination from sodium, calcium, magnesium, and other divalent cations was obtained in this step. Subsequently, the americium and plutonium were eluted either upflow or downflow with about six column volumes of $7\underline{M}$ HNO_3 to yield a product solution containing 2 to 4 g/L each of [241]Am and plutonium. Final purification of the [241]Am recovered in the Hanford PRF was accomplished in Pacific Northwest Laboratory facilities using ion exchange displacement chromatographic technology developed by Wheelwright.([12]) Americium and plutonium in the $7\underline{M}$ HNO_3 solution obtained from the previous cation exchange concentration step were separated by sorption of the plutonium on Dowex 1 (Dow Chemical Co.) anion exchange resin. Subsequently, the 7$\underline{M}$ HNO$_3$ waste stream containing the ^{241}Am was diluted with water to yield a 1$\underline{M}$ HNO$_3$ solution containing 0.25 to 0.5 g ^{241}Am/L. Americium in this feed was loaded onto a 10.8-cm diameter bed of H$^+$-form Dowex 50-X8 (50-100 mesh) cation exchange resin and then eluted through a series of four Zn^{2+}-form Dowex 50-X8 resin beds with a 0.105$\underline{M}$ nitrilotriacetic acid solution buffered to pH 6.5 with NH$_4$OH. Displacement elution was performed at 60°C at a flowrate of 8 mL/(cm^2)(min). The center product cut from the final resin bed contained 8 to 9 g/L of highly purified ^{241}Am. Oxalic acid was added to this latter solution, and the resulting oxalate precipitate was calcined to AmO$_2$.

Ritter and Bray(13) proposed a batch solvent extraction-strip purification scheme (PAMEX process) as an alternative to the chromatographic process. Feed for the extraction step in the PAMEX process is prepared by making the crude americium concentrate from the cation exchange step 0.5$\underline{M}$ hydroxyacetic acid (HOAc) and 0.09$\underline{M}$ diethylenetriaminepentaacetic acid (DTPA) and then adjusting its pH to 1-2. The HOAc is a buffering agent while DTPA is added to complex and suppress extraction of impurities (e.g., Ca^{2+}, Al^{3+}, etc.). Contact of the adjusted aqueous feed with an equal volume of 0.4$\underline{M}$ HDEHP (bis-2ethylhexylphosphoric acid)-0.2$\underline{M}$ TBP-kerosene at 25°C extracts 90-95% of the americium. Americium is stripped from the HDEHP phase with one-quarter volume of 0.5$\underline{M}$ HOAc-0.09$\underline{M}$ DTPA-pH 3.3 solution. Addition of oxalic acid to the strip solution precipitates Am$_2$(C$_2$O$_4$)$_3$ which can be calcined to AmO$_2$. Although successfully tested on a laboratory-scale, the PAMEX process was never used on a plant-scale primarily because the chromatographic ion exchange process was already in operation and satisfactorily producing high-purity ^{241}Am.

Other Extractants. Historically, development of the DBBP extraction process benefited much from the work of Walsh(14) at the Los Alamos Scientific Laboratory; Walsh developed a process for TBP extraction of Am(III) from low-acid high-salt solutions.

Limited laboratory tests to explore the feasibility of substituting TOPO (trioctylphosphine oxide) for DBBP were performed. Advantages cited by Bruns(15) for use of 0.15$\underline{M}$ TOPO-CCl for DBBP in the americium extraction scheme would not eliminate the need for a difficult-to-control feed acidity adjustment step. Primarily for this latter reason, further consideration of TOPO as an alternative americium extractant was not actively pursued.

Schulz(16) during 1973-1977 conducted very extensive laboratory-scale tests of the use of 30 vol% DHDECMP (dihexyl-N, N-diethylcarbamoyl-methylene phosphonate)-CCl_4 to extract Am(III) and Pu(IV) from high-acid CAW solution. Whereas monodentate DBBP molecules contain only an active phosphoryl group, bidentate DHDECMP molecules contain both active phosphoryl and carbonyl groups. By virtue of their bidentate characteristics, DHDECMP solutions are able to extract trivalent americium very well from concentrated HNO_3 solutions (Figure 8). Substitution of 30 vol% DHDECMP for 30 vol% DBBP in PRF waste treatment operations permits direct extraction of both Am(III) and Pu(IV) from acid CAW solution and completely eliminates problems associated with addition of concentrated NaOH to the CAW solution. Substitution of DHDECMP for DBBP does not require any major flowsheet changes since dilute HNO_3 and HNO_3-HF solutions can still be used, respectively, to selectively partition Am(III) from Pu(IV) and to strip residual americium and plutonium. However, because the DHDECMP solvent extracts considerable HNO_3, substitution of thermal concentration for ion exchange concentration of the ^{241}Am product would be desirable.

In addition to collection of comprehensive distribution data, Schulz's investigations embraced ways of purifying commercial-grade DHDECMP, alpha radiolysis of DHDECMP-CCl_4 solvents, and mixer-settler flowsheet tests with both synthetic and actual CAW solutions. Collectively, results of this research clearly demonstrated the many advantages of substituting DHDECMP for DBBP. Shutdown of PRF Am(III) extraction operation in 1976 prevented follow-on plant tests of the DHDECMP flowsheet.

McIsaac and coworkers(17,18) have recently developed and tested chemical flowsheets for DHDECMP extraction of trivalent americium and other actinides from the high-acid waste solution produced in Idaho Chemical Processing Plant operations. In cooperation with McIsaac, Oak Ridge National Laboratory scientists, Tedder, Blomeke, and Bond,(19) have also demonstrated DHDECMP extraction technology for removal of actinides, including ^{241}Am and ^{243}Am, from Purex process acid waste. These research and development efforts have convincingly established the superiority of DHDECMP over all other currently known reagents for engineering-scale extraction of Am^{3+} from strong HNO_3 solutions.

Summary

Plant-scale DBBP solvent extraction facilities were operated at Hanford for about six years to recover ^{241}Am from an acid (2M HNO_3) waste stream. Although several kilograms of americium were recovered and partially purified, overall plant performance

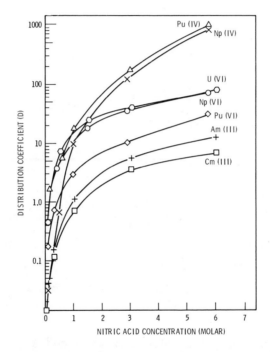

Figure 8. Extraction of +3, +4, and +6 actinides from HNO₃ solutions by 30% DHDECMP–diisopropylbenzene solvent

and experience was only marginal. Both equipment and process deficiencies contributed to lower (66-80%) than desired (95-100%) product recoveries. In particular, major process difficulties and upsets were encountered in conducting in-line addition of concentrated NaOH to an unbuffered solution to adjust feed acidity to the desired $0.1\underline{M}$ HNO_3. Substitution of the bidentate extractant DHDECMP for DBBP to permit extraction of Am^{3+} directly from the $2\underline{M}$ HNO_3 solution was shown to be feasible in bench scale tests; shutdown of americium recovery operations at Hanford occurred before plant tests of the DHDECMP extractant could be made.

Literature Cited

1. Smith, A. E.; Compiler, U.S. Atomic Energy Commission Report ARH-2915, Atlantic Richfield Hanford Company, Richland, Washington, 1973.

2. Groot, C.; and Hopkins, H. H., Jr., U.S. Atomic Energy Commission Report HW-23344, General Electric Company, Richland, Washington, 1952.

3. Bruns, L. E.; Chemical Engineering Progress Symposium Ser., 1967, 63, 156.

4. Bruns, L. E.; "Plutonium - Uranium Partitioning by a Reflux Extraction Flowsheet," in "Proceedings, ISEC 71," Society of Chemical Industry, London, 1971, Vol. 1, p. 186.

5. Zemlyanukhin, G. P.; Savoskina, G. P.; and Pushlenkov, M. F.; Radiokhimiya, 1962, 4, 570.

6. Sheppard, J. C.; U.S. Atomic Energy Commission Report HW-81166, General Electric Company, Richland, Washington, 1964.

7. Kingsley, R. S.; U.S. Atomic Energy Commission Report RL-SEP-518, General Electric Company, Richland, Washington, 1965.

8. Siddall, T. H., Jr.; U.S. Atomic Energy Commission Report DP-219, E. I. duPont de Nemours Company, Savannah River Laboratory, Aiken, South Carolina, 1957.

9. Richardson, G. L.; U.S. Atomic Energy Commission Report BNWL-CC-1503, Pacific Northwest Laboratory, Richland, Washington, 1968.

10. Taylor, I. N., Jr.; U.S. Atomic Energy Commission Report ARH-210, Atlantic Richfield Hanford Company, Richland, Washington, 1967.

11. Hammelman, J. E.; U.S. Atomic Energy Commission Report ARH-2824, Atlantic Richfield Hanford Company, Richland, Washington, 1973.

12. Wheelwright, E. J.; "Kilogram - Scale Purification of Americium by Ion Exchange," paper presented at Symposium on Separation Science and Technology for Energy Application, Gatlinburg, Tennessee, October 30 - November 2, 1979.

13. Ritter, G. L.; and Bray, L. A.; U.S. Atomic Energy Commission Report ISO-95, Isochem. Corp., Richland, Washington, 1966.

14. Walsh, K. A.; U.S. Atomic Energy Commission Report LA-1861, Los Alamos Scientific Laboratory, Los Alamos, New Mexico, 1955.

15. Bruns, L. E.; U.S. Atomic Energy Commission Report ARH-2426, Atlantic Richfield Hanford Company, Richland, Washington, 1972.

16. Schulz, W. W.; U.S. Atomic Energy Commission Report ARH-SA-203, Atlantic Richfield Hanford Company, Richland, Washington, 1973.

17. McIsaac, L. D.; Baker, J. D.; and Tkachyk, J. W.; U.S. Energy Research and Development Report ICP-1080, Allied Chemical Company, Idaho Falls, Idaho, 1975.

18. McIsaac, L. D.; and Schulz, W. W.; Removal of Actinides from Nuclear Fuel Reprocessing Waste Solution with Bidentate Organophosphorus Extractants in "Transplutonium 1975," Muller, W. and Lindner, R., eds., North Holland Pub. Company, Amsterdam, 1976.

19. Tedder, D. W.; and Blomeke, J. O.; U.S. Department of Energy Report ORNL/TM-6480, Oak Ridge National Laboratory, Oak Ridge, Tennessee, 1978.

RECEIVED December 24, 1980.

TRU FACILITY
PROCESSES AND EXPERIENCE

8

Experience in the Separation and Purification of Transplutonium Elements in the Transuranium Processing Plant at Oak Ridge National Laboratory

L. J. KING, J. E. BIGELOW, and E. D. COLLINS

Oak Ridge National Laboratory, Oak Ridge, TN 37830

The Transuranium Processing Plant (TRU) (1,2,3) at Oak Ridge National Laboratory (ORNL) is the production, storage and distribution center for the heavy-element research program of the U. S. Department of Energy (DOE) and its predecessors, the U. S. Energy Research and Development Administration and the U. S. Atomic Energy Commission. TRU and the neighboring High Flux Isotope Reactor (HFIR) were built to produce quantities of the transuranium elements for use in research. Operations in both facilities were begun in 1966. Since then, TRU has been the main center of production for transcurium elements in the United States, producing 460 mg of ^{249}Bk, 4 g of ^{252}Cf, 18 mg of ^{253}Es, and 10 pg of ^{257}Fm.

Target rods containing plutonium, americium, and curium are remotely fabricated at TRU, irradiated in the HFIR, and then processed at TRU for the separation and purification of the heavy actinide elements. All elements from plutonium through fermium are separated and purified. Portions of the plutonium, americium, and curium are refabricated into targets for additional irradiation. The berkelium, californium, einsteinium, and fermium are distributed to researchers. More than 1000 shipments of these materials have been made to about 30 different laboratories in the United States and several foreign countries. The complete production history of key isotopes is given in Table I. During 14 years of operation, 39 chemical processing campaigns have been completed at TRU to process about 265 targets which had been irradiated in the HFIR and about 195 targets which had been irradiated in a reactor at the Savannah River Plant (SRP).

There are three distinct periods in the history of TRU operations. From 1966 to 1970, TRU had been recovering transuranium elements as rapidly as they could be produced in the HFIR and fabricating targets as fast as feed material was being recovered for recycle. Production rates were limited because the feed materials that were available (^{242}Pu and curium containing predominantly ^{244}Cm) require long irradiation periods (up to 18 months) to produce appreciable amounts of the heavier elements.

0097-6156/81/0161-0133$05.00/0
© 1981 American Chemical Society

In the period from 1970 through 1973, in addition to the processing of irradiated HFIR   targets, operations at TRU were
expanded to include the processing of special SRP targets that
had been irradiated as part of the Californium-I campaign, an
irradiation and processing campaign to provide ^{252}Cf for use in
an SRP program to evaluate the commercial market for ^{252}Cf.
Approximately 720 mg of ^{252}Cf was recovered for the SRP program,
and 94 mg of ^{249}Bk and 5 μg of ^{254}Es recovered from the SRP
targets was used in the research programs. Then, beginning in
1974, TRU operations reverted to the processing of targets fabricated at TRU and irradiated in the HFIR. However, since then,
the rates of production of the transcurium isotopes have been
considerably higher than in earlier years because some of the
curium recovered from the Californium-I material has been used as
feed for the HFIR targets. Berkelium, californium, einsteinium,
and fermium are produced much more rapidly from this curium
because it is rich in the heavier isotopes (curium-246-248) which
can be transmuted to transcurium isotopes in shorter irradiation
periods.

TABLE I. KEY ISOTOPE PRODUCTION HISTORY AT TRU

Fiscal Year	^{242}Pu (g)	^{243}Am (g)	^{244}Cm (g)	^{249}Bk (mg)	^{252}Cf (mg)	^{253}Es (mg)	^{257}Fm (pg)
1967	87	25	134	0.34	5.6	0.014	0
1968	0	188	212	0.05	0.5	0	0
1969	15	5	57	2.2	15	0.1	0.07
1970	8	13	72	7.6	52	0.4	0.19
1971	10	3	439	37	284	0.7	0.72
1972	16	3	350	66	513	0.9	0.85
1973	5	4	240	49	428	1.6	1.25
1974	0	2	87	39	386	2.2	1.5
1975	0	3	104	75	717	3.8	1.6
1976	0	2	50	29	277	1.7	0.54
1977	0	0	38	52	499	2.6	1.1
1978	0	0.6	48	67	632	3.6	1.6
1979	0	3	21	32	322	0.8	0.6
Totals	141	252	1852	456	4131	18.4	10

Facilities and Equipment

The heart of TRU is a battery of nine heavily shielded hot
cells housed in a two-story building. Of the nine cells, four
contain chemical processing equipment for dissolution, solvent
extraction, ion exchange, and precipitation operations. Three

contain equipment for the preparation and inspection of HFIR
targets, and two cells are used for analytical chemistry
operations. In addition, there are eight laboratories (four on
each floor) used for process development, for process-control
analyses, and for final product purification and packaging
operations.

Figure 1 shows a cross section of a typical cell and the
surrounding building areas. The top and back of the cell line is
enclosed by a high-bay area (the "limited access area") that is
equipped with a bridge crane. Removable top plugs provide access
to the cells. Service lines enter the cells through removable
plugs in the back walls and tops. Shielded pits in the floor
behind the cells house off-gas filters and a pipe tunnel for process lines. The front face of the cell line makes up one wall of
the operating area. Essentially all process and building service
instrumentation is located in the operating area. The second
floor immediately over the operating area is a chemical makeup
area for process-reagent head tanks, uncontaminated pumps, etc.
Transmitters for process and service instrumentation are located
in this area.

Within each shielded cell, process equipment is enclosed in
a fixed containment box (the cell cubicle), which is about a two
meter cube. Small items of chemical processing equipment, such
as valves, pumps, ion exchange columns, and solution sampling
devices, are mounted on racks in the cubicles. This equipment
can be installed or removed remotely by using manipulators and
air-operated impact wrenches. Contaminated equipment can be
introduced into or removed from the cell cubicles through an
intercell conveyor to a glove box or to a shielded carrier at a
loading station at one end of the cell bank. Alternatively,
equipment may be introduced or removed through the top of the
cubicle by use of a shielded caisson (transfer case) designed to
maintain shielding and contamination control during the transfer.

A tank pit for housing waste collection equipment and
process and storage tanks is located behind and below each cell
cubicle and shielded from it by a concrete wall. The equipment
in the tank pits is serviced and maintained by means of a combination of contact, remote, and underwater maintenance techniques.

The extensive provisions at TRU for changing and modifying
equipment have allowed continual updating of the plant to
include new concepts in chemical processes (4) and equipment
design (5). Since the beginning of operation, the processing
equipment in the cubicles has effectively been changed twice
through replacement of 17 equipment racks in 9 cell positions.
The flexibility and reliability of the chemical processing
equipment and techniques have been improved significantly through
this evolutionary process (6).

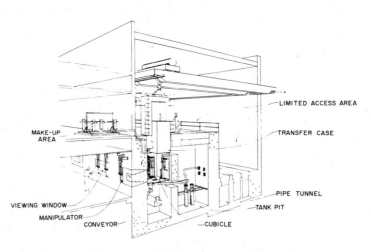

MAKE-UP
AREA

VIEWING WINDOW
MANIPULATOR
CONVEYOR

LIMITED ACCESS AREA

TRANSFER CASE

PIPE TUNNEL
TANK PIT
CUBICLE

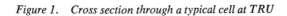

TRANSURANIUM PROCESSING PLANT

Figure 1.   Cross section through a typical cell at TRU

Transuranium Element Chemical Processing

Since operation of TRU began, numerous process changes have been made to counteract problems caused by equipment corrosion, to satisfy new processing requirements, and to utilize improved processes. In initial operations at TRU, unexpected corrosion of the Zircaloy-2 equipment occurred in high specific-activity chloride solutions, and the dissolved zirconium caused operating problems that made some of the flowsheets for the processing steps either difficult to operate or totally unusable. Because of these and other problems, several new flowsheets were devised or adopted, tested, and scaled to plant-size equipment.

Chemical processing to recover transuranium elements is accomplished in a series of separate steps called a processing campaign. In a typical campaign, about ten irradiated HFIR targets plus rework material from previous campaigns are processed over a period lasting about two months. The composition of the feed material is shown in Table II. Of the transplutonium elements in the feed, approximately 25% of the curium and 5% of the berkelium and californium are rework material from the previous campaign. In addition to the components shown, gram quantities of zirconium, nickel, iron, and chromium are frequently present from equipment corrosion and must be removed.

TABLE II.
APPROXIMATE COMPOSITION OF TYPICAL TRU CAMPAIGN FEED:
TEN HFIR TARGETS PLUS REWORK MATERIAL

Component	Source	Weight (g)
Al	Target cladding, spacers, pellet matrix	1200
Cm[a]	Target residual, plus rework	65
Si	Activation of aluminum	25
Mo, Ru, Pd, Cs, Ba	Fission products	25[b]
Rare-Earth Elements	Fission products	15[b]
Ni	Impurity in aluminum	10
Zn	Impurity in aluminum	7
Fe	Impurity in aluminum	7
Mn, Cu, Mg, Cr, Ti	Impurity in aluminum	5[b]

[a]Other transplutonium elements in the feed include 1 g of Am, 30 mg of Bk, 300 mg of Cf, 1 mg of Es, and 1 pg of Fm.
[b]Total amount from the group.

The sequence of steps that is now being used successfully to process HFIR targets in the mainline cell bank is shown in Fig. 2. The processing steps generally tend to separate the

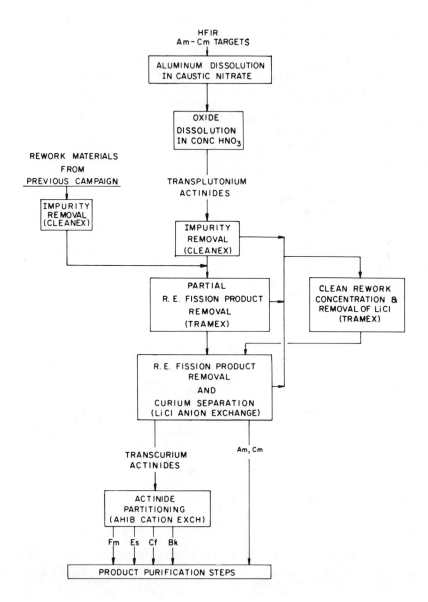

Figure 2.   Sequence of mainline steps used to process HFIR targets

transplutonium elements as a group from fission products and gross impurities before partitioning and purifying the elements. Wastes from the processing are transferred to the ORNL Intermediate Level Waste system for subsequent treatment and disposal.

Target Dissolution. A HFIR target is a 9.4-mm-diam by 0.89-m-long rod encased within a cylindrical tube which serves to channel the flow of cooling water in the reactor. The coolant-flow tubes are mechanically removed from irradiated target rods before dissolution because the rate of dissolution of tubes is more difficult to control than that of rods.

A well-known, two-part method (7) is used for dissolution of targets. The aluminum is preferentially dissolved in caustic-nitrate solution; and after the aluminum-bearing solution is removed, the residual oxides (transuranium elements and fission products) are dissolved in concentrated HNO_3. The group of target rods is put into a water-jacketed Zircaloy-2 tank and covered with a solution of 2.1 $\underline{M}$ $NaNO_3$ (using a $NaNO_3$-to-aluminum mole ratio of 1.2), which is then heated and held at 90 to 94°C while 10 $\underline{M}$ NaOH is added. The reaction is highly exothermic and the rate of dissolution, as indicated by the temperature difference between the solution in the dissolver tank and the water jacket, is controlled within empirically determined limits by varying the rate of addition of 10 $\underline{M}$ NaOH. The average dissolution rate is about 10 mol of aluminum per hour. Addition of NaOH is continued until a ratio of 2.4 mol of NaOH per mole of aluminum is reached. Approximately 25 liters of $NaNO_3$ solution and 10.8 liters of NaOH solution are required for 10 targets. Calculations made from laboratory data indicate that for each mole of aluminum dissolved, 0.37 mol of NH_3 and 0.02 mol of H_2 are evolved into the dissolver off-gas. These gases are diluted to nonexplosive concentrations by an air purge of the dissolver tank. The aluminum-bearing caustic-nitrate solution is pumped from the dissolver to a holding tank through a porous stainless steel filter. About 0.1% of the actinides are lost to the caustic-nitrate solution.

Concentrated HNO_3 and water are added to the dissolver vessel to comprise about 7 liters of 5.5 $\underline{M}$ HNO_3 which is simmered at 103°C for two hours to dissolve the actinides.

A major product of a processing campaign is ^{253}Es, which has a half life of 20.5 d. Thus, it is important to process HFIR targets after only short decay. Because of this the targets usually contain about 7 TBq of ^{131}I at the start of a campaign.

About one-half of the radioiodine in the targets remains in the caustic-nitrate solution which is stabilized with sodium thiosulfate and stored until the radioiodine decays. During the acid dissolutions, the dissolver off-gas is directed to the iodine sorption system in which the gas is contacted with a 10 $\underline{M}$ HNO_3--0.4 $\underline{M}$ $Hg(NO_3)_2$ solution in an absorption column. Then, the

iodine is evolved from the dissolver solution by simmering, air
sparging, and H_2O_2 additions until the solution contains less
than 0.2 TBq of ^{131}I. Iodine is removed from the off-gas by a
factor of about 500. The mercuric nitrate solution is neutralized and stored until the radioiodine decays and is then
disposed of to the Intermediate Level Waste system.

Miscellaneous Impurity Removal. The dissolver solution is
treated by means of a two-stage Cleanex (8) batch solvent extraction to remove miscellaneous metallic impurities. The nitric
acid dissolver solution is concentrated to a small volume ($\sim$2L)
to remove excess acid, HCl is added, and the residue is digested
to dissolve zirconium precipitates. The solution is then diluted
to about 12 L and adjusted to about 0.25 $\underline{M}$ acidity with NaOH. An
oxidant, 1.5 $\underline{M}$ LiOCl, is added to bring the concentration to 0.1
$\underline{M}$ LiOCl to oxidize any molybdenum in the feed to the extractable
$\overline{Mo}$(VI) form. The adjusted feed solution is then contacted with
an organic solution of about 35 liters of 1 $\underline{M}$ bis-2-ethylhexyl
hydrogen phosphate (HDEHP) in normal paraffin diluent to extract
the transuranium elements. During extraction, 5 $\underline{M}$ NaOH is added
to adjust the aqueous solution to about 0.03 $\underline{M}$ acidity to enhance
the extraction coefficient. The aqueous phase is transferred to
another tank where a second cycle of extraction is used to
recover residual transuranium elements. Five liters of 1.6 $\underline{M}$
Adogen (an 8- to 10-C tertiary amine) in diethylbenzene (DEB)
diluent is added to hold iron in the organic phase during
stripping, and the transuranium elements are stripped from the
organic extract using a 6 $\underline{M}$ HCl--0.5 $\underline{M}$ H_2O_2 solution.

Rare-Earth Removal and Separation of Curium from Heavier
Elements. The Cleanex product solution is combined with rework
material from the previous campaign and the composite solution
is processed by means of a Tramex (9) batch solvent extraction to
obtain a product containing about 98% of the transcurium
elements, 90% of the curium, and 10% of the rare earths. The
batch Tramex process consists of the following: (a) adjusting
the composite solution to about 10 liters of 12 $\underline{M}$ LiCl--0.04 $\underline{M}$
HCl; (b) extracting all transplutonium and rare-earth elements
with about 20 liters of 1 $\underline{M}$ Adogen in diisopropylbenzene (DIPB);
(c) transferring the raffinate to another tank for rework
processing; (d) adding about 20 liters of a solution of 0.6 $\underline{M}$
Adogen--0.03 $\underline{M}$ HCl--0.05 $\underline{M}$ dibutylhydroquinone in DEB to dilute
the organic phase composition to 0.8 $\underline{M}$ Adogen; (e) scrubbing out
10% of the curium and 90% of the rare earths with 10.6 $\underline{M}$
LiCl--0.02 $\underline{M}$ HCl; and (f) stripping the organic extract with 8 $\underline{M}$
HCl--0.1 $\underline{M}$ H_2O_2 to recover the transplutonium elements. The
scrub raffinate is stored temporarily and processed later with
other clean rework to recover the actinides from the rare earths.
    In preparation for subsequent anion-exchange runs, the
Tramex product solution is treated by a two-step clarification

process to eliminate problems with solids formation and to significantly reduce the amount of actinide elements diverted into rework solutions. In the first step, the Tramex product solution is washed with DEB to remove entrained Adogen which would be degraded in the subsequent boiling step to a tar that would sorb a significant amount of the transplutonium elements. In the second step, treatment of the Tramex product solution includes: (a) adjustment to 12 $\underline{M}$ LiCl--1 $\underline{M}$ HCl; (b) filtration to remove insoluble materials such as aluminum, zirconium, and sodium; (c) dilution by flushing the filter; and (d) readjustment to about 5 liters of 12 $\underline{M}$ LiCl--0.1 $\underline{M}$ HCl to provide a clear feed solution for the anion-exchange runs.

The clarified Tramex product solution is divided into two or three batches ($\leq$35 g of curium or $\leq$19 g of ^{244}Cm per batch) and processed by LiCl-based anion exchange, which is discussed in detail in another paper at this symposium (10), to obtain further decontamination from rare earths and to separate curium from the heavier elements. In each run, the transplutonium and rare-earth elements are sorbed on Dowex 1-X10 ion exchange resin from a 12 $\underline{M}$ LiCl solution. Rare earths are eluted with 10 $\underline{M}$ LiCl, curium with 9 $\underline{M}$ LiCl, and the transcurium elements with 8 $\underline{M}$ HCl. About 5% of the curium is purposely eluted along with the transcurium elements to prevent losses of ^{249}Bk, which elutes immediately after the curium and is not distinguishable by the in-line instrumentation. The transcurium element fractions from each run are combined and processed in a second-cycle run, using new resin, to remove most of the excess curium.

Transcurium Element Separation. The transcurium elements are separated by means of a high pressure ion-exchange process, which is described in another paper at this symposium (11). Feed for this process (0.25 $\underline{M}$ HNO$_3$) is prepared by precipitating the transcurium element product from the second-cycle anion-exchange run with LiOH, filtering to separate the precipitated transcurium element hydroxides from the LiCl-bearing solution, and dissolving the filtered precipitate in HNO$_3$. About half of the feed solution ($\leq$200 mg of ^{252}Cf per batch) is processed in each of two high-pressure ion-exchange runs. In each run, the transcurium elements are loaded onto Dowex 50W-X8 resin in a short "loading" column (200 mm long) and then chromatographically eluted through a longer (1.2 m) column. This technique reduces radiation damage to the resin in the long column and enables reuse in several runs. The resin in the short column is replaced after each run. The fermium, einsteinium, and californium are eluted with 0.25 $\underline{M}$ ammonium alphahydroxyisobutyrate (AHIB) at pH 4.2, the berkelium with 0.25 $\underline{M}$ AHIB (pH 4.6), and the residual curium with 0.50 $\underline{M}$ AHIB (pH 4.8). Product yields are usually greater than 90%, and excellent separations are obtained. The berkelium is decontaminated from californium by a factor of 500, and the einsteinium is decontaminated from californium by a factor of 10^3 to 10^4.

Rework Processing. Immediately following a campaign, all clean solutions containing significant amounts of the transplutonium elements (usually about 10% of each element in the campaign feed) are accumulated and reprocessed. "Clean" solutions are those that have a known history, such as raffinates from process steps and solutions which are used to flush the equipment internally following the campaign. These solutions do not contain a large amount of corrosion products or other potentially troublesome components.

"Dirty" rework materials are obtained from the product purification steps, target fabrication, and equipment flushing. Since numerous temporary piping connections are made to accommodate multipurpose use of several tanks during a campaign, leaks and/or small spills can occur. Thus, the cell cubicle floors and external portions of the equipment are flushed thoroughly after each campaign. All of the rework materials obtained between campaigns are accumulated, processed by means of one or two cycles of the Cleanex batch extraction process to remove miscellaneous impurities, and recycled to the next campaign.

Product Finishing. The sequences of steps used to purify each of the products are shown in Fig. 3. A three-step sequence is used to purify the americium-curium (referred to hereafter as "curium" since very little americium is present normally) and convert this material to the oxide form for use in HFIR targets. The curium product solutions from each of the first cycle LiCl-based anion-exchange runs are combined and processed by means of a Tramex batch extraction to remove most of the LiCl. After the Tramex product is converted to a nitrate medium by adding HNO_3 and evaporating HCl, the curium solution is divided into batches, each containing 25 g of curium or less, and each batch is purified from miscellaneous impurities (predominantly the residual lithium) by means of two cycles of oxalate precipitation. The curium oxalate is then treated by means of a boiling HNO_3 technique to decompose oxalic acid. Following this, the curium oxide is prepared by a cation resin loading, calcination technique, which is described in another paper (12) at this symposium. The impurities remaining in a typical batch of curium oxide are all below the calculated limits for heat production and neutron absorption in the HFIR.

The berkelium product fractions from the transcurium element separation runs are combined and the composite solution is concentrated and purified (13) from AHIB solution by sorbing the actinides on a cation exchange resin and stripping with 8 $\underline{M}$ HNO_3. The berkelium is then purified from ^{252}Cf by means of one or more cycles of batch solvent extraction (Berkex) to reduce the ^{252}Cf content to less than 0.25 µg (which requires a californium DF of about 3000). $NaBrO_3$ is added (to 0.3 $\underline{M}$) to oxidize Bk(III) to Bk(IV), which is extracted into a solution of 0.5 $\underline{M}$ HDEHP in dodecane diluent. The organic and aqueous phases are separated and the berkelium is stripped into 8 $\underline{M}$ HNO_3--1 $\underline{M}$ H_2O_2.

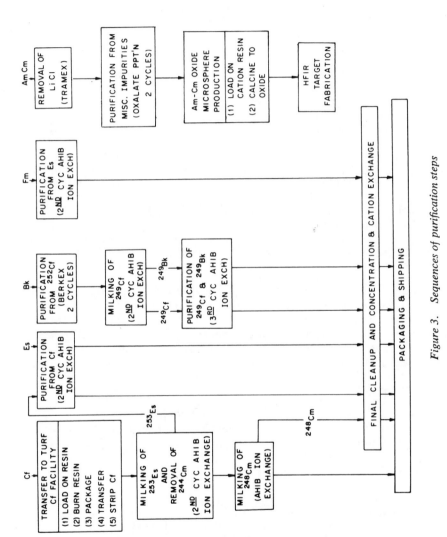

Figure 3. Sequences of purification steps

The berkelium, californium, einsteinium, and fermium products are then packaged and transferred from the main cell bank to other facilities in which they are purified further. The californium is sorbed on about 0.9 mL of cation exchange resin in a platinum column, which is then calcined at 650°C (923 K) to convert the californium to the oxysulfate. (In this form, the californium can be stored for long periods and then be recovered easily in a few milliliters of HNO_3.) Then, the platinum column is put into a special package and is transferred by a pneumatic transfer system (14) to a californium purification facility in a hot cell in the adjacent Thorium Uranium Recycle Facility (TURF). A separate facility is required to minimize ^{244}Cm contamination and to permit subsequent recovery in high isotopic purity of the ^{248}Cm daughter of ^{252}Cf.

The normal processing sequence is to store the californium product about one month to permit the ^{253}Cf (a minor constituent of the californium) to decay into isotopically pure ^{253}Es. Then the package containing the column is connected to an equipment rack, and the californium is leached from the package. More than 99.8% of the californium is readily dissolved and removed from this package with about 0.1 liter of 0.5 $\underline{M}$ HNO_3.

The isotopically pure ^{253}Es is recovered and the californium is highly purified from curium by high-pressure ion exchange using AHIB. The californium is loaded into another platinum ion exchange column and stored pending subsequent processing to recover the ^{248}Cm daughter product.

The fermium, einsteinium, and berkelium are transferred from the main hot cells and are purified further and prepared for shipment in a small hot cell and in glove box facilities that are kept free from undesirable contaminants. The chemical processing (13) involves numerous additional cycles of ion exchange purification on the micro scale.

Acknowledgement

This research was sponsored by the Office of Basic Energy Sciences, U. S. Department of Energy, under contract W-7405-eng-26 with the Union Carbide Corporation.

Literature Cited

1. Ferguson, D. E.; Nucl. Sci. Eng., 17, 1963, 435. Ten following articles describe details of the program.
2. King, L. J.(compiler); Safety Analysis for the Transuranium Processing Plant, Building 7920, ORNL-3954 (April 1968).
3. King, L. J. and Matherne, J. L.; Proc. 14th Conf. Remote Syst. Technol., 1966, 21-27.
4. Collins, E .D. and Bigelow, J. E.; Proc. 24th Conf. Remote Syst. Technol., 1976, 130-139.

5. Chattin, F. R.; King, L. J.; Peishel, F. L.; Proc. 24th Conf. Remote Syst. Technol., 1976, 118-129.
6. King, L. J.; Proc. 27th Conf. Remote Syst. Technol., 1979, 96-101.
7. Martens, R. I.; Poe, W. L. Jr.; and Wible, A. E.; Chem. Engr. Progr. Symp. Ser. No. 51, 60, 1964, 44.
8. Bigelow, J. E.; Collins, E. D.; and King, L. J.; Actinides Separations, ACS Symp. Series, 117, 1979, 147-155.
9. Bigelow, J. E.; Chattin, F. R.; and Vaughen, V. C. A.; Proc. Int. Solvent Etr. Conf., I, 1971, 507.
10. Collins, E. D.; Benker, D. E.; Chattin, F. R.; Ore, P. B.; Ross, R. G., "Multigram Group Separation of Actinide and Lanthonide Elements by LiCl-Based Ion Exchange", paper presented at Symposium on Industrial-Scale Production-Separation-Recovery of Transplutonium Elements, 2nd Chem. Congr. North American Continent, Las Vegas, NV, 1980.
11. Benker, D. E.; Chattin, F. R.; Collins, E. D.; Knauer, J. B.; Orr, P. B.; Ross, R. G.; Wiggins, J. T.; "Chromatographic Cation Exchange Separation of Decigram Quantities of Californium and Other Transplutonium Elements", paper presented at Symposium on Industrial Scale Production-Separation-Recovery of Transplutonium Elements, 2nd Chem. Congr. North American Continent, Las Vegas, NV, 1980.
12. Chattin, F. R.; Benker, D. E.; Lloyd, M. H.; Orr, P. B., Ross, R. G.; Wiggins, J. T.; "Preparation of Curium-Americium Oxide Microspheres by Resin-Bead Loading", paper presented at Symposium on Industrial Scale Production-Separation-Recovery of Transplutonium Elements, 2nd Chem. Congr. North American Continent, Las Vegas, NV, 1980.
13. Baybarz, R. D.; Knauer, J. B.; Orr, P. B.; Final Isolation of the Transplutonium Elements from the Twelve Campaigns Conducted at TRU During the Period August 1967-December 1971, ORNL-4672 (April 1973).
14. Peishel, F. L.; Burch, W. D.; Jarvis, J. P.; Proc. 18th Conf. Remote Syst. Technol., 1980, 93-100.

RECEIVED December 19, 1980.

Multigram Group Separation of Actinide and Lanthanide Elements by LiCl-Based Anion Exchange

E. D. COLLINS, D. E. BENKER, F. R. CHATTIN,
P. B. ORR, and R. G. ROSS

Oak Ridge National Laboratory, Oak Ridge, TN 37830

A chromatographically operated, LiCl-based anion exchange
(LiCl AIX) process (1) was adapted from the laboratory scale to
the multigram scale and has been used successfully in the Transuranium Processing Plant (TRU) at Oak Ridge National Laboratory
(ORNL) for over ten years to separate lanthanide fission products
from the transplutonium actinides and to partition americium and
curium from the heavier elements. During early process design
studies for TRU, the LiCl AIX process was recognized as one of
the few methods that had been used successfully in laboratory
operations; however, the use of solid ion exchangers was considered impractical for the larger-scale operations at TRU, primarily
because of the localized heating and radiolytic gas generation that
would occur. Thus, a continuously operated solvent extraction
process (Tramex) was developed to accomplish the actinide-
lanthanide separation (2). This process was chemically similar
to the anion exchange process because it utilized a mixed tertiary amine (primarily octyl and decyl) to extract transplutonium
actinides from a concentrated LiCl solution.
   A second solvent extraction process (Pharex) was developed
to partition the transcurium actinides from the americium and
curium in the Tramex product (3). The Pharex process utilized
2-ethylhexyl phenylphosphonic acid as the extractant for the
transcurium actinides. During early operations, the selectivity
of the Pharex extractant was found to be severely reduced by the
presence of zirconium ions, which were introduced into the process
solutions by corrosion of Zircaloy-2 equipment in TRU. At zirconium concentrations above 10 ppm, the achievable separation
began to be diminished and, at 100 ppm, a practical separation
could not be made (4). Thus, a replacement for the Pharex process was needed, and the LiCl AIX process was the most immediate
alternative.
   Temporary glass equipment was installed, and the LiCl AIX
process was successfully scaled to a useful level (5). Tramex
product solutions containing from 4 to 10 g of ^{244}Cm (11 to 28 W
of decay heat) were processed initially, using a 38-mm-diam
column containing 450 mL of Dowex 1-X8 anion exchange resin.

0097-6156/81/0161-0147$05.00/0
© 1981 American Chemical Society

Dowex 1-X10 resin is now used. Subsequently, a larger glass
column, having a diameter of 50 mm and containing 1.2 L of resin,
was used. Finally, the large glass column was replaced with a
tantalum column of identical size. A loading capacity of 19 g of
^{244}Cm (54 W) or 35 g of total actinide mass has been empirically
established for the tantalum column. Localized heating and cumu-
lative radiation exposure of the resin are problems, although not
as extensive as expected. Radiolytic gas generation has not
caused any significant difficulty, and downflow operation is used
with little evidence of channeling. At the loading limits that
have been established, three column loadings and elutions can be
made successfully on each batch of resin.

Process Chemistry

The LiCl AIX process is based on (i) the formation of
anionic chloride complexes of the tripositive actinide and lanthanide metals in concentrated LiCl solutions, (ii) the sorption
of these complexes onto a strong base anion exchange resin contained in a column, and (iii) the preferential chromatographic
elution of the lanthanides as a group prior to elution of the
actinides. The generalized formation of the trivalent metal
anionic chloride complexes is illustrated in equation (1).

$$M^{3+}_{aq} + xCl^- \rightleftharpoons MCl_x^{3-x} . \qquad x > 3 \qquad\qquad (1)$$

There is no evidence of the structure of the species or group of
species that is sorbed by the resin. However, for purposes of
illustration, the formation and sorption of the divalent anionic
complex are shown in equations (2) and (3):

$$M^{3+}_{aq} + 5Cl^-_{aq} \rightleftharpoons MCl_5{}^{2-}_{aq} , \qquad\qquad (2)$$

$$2R_4NCl_{org} + MCl_5{}^{2-}_{aq} \rightleftharpoons (R_4N)_2MCl_{5org} + 2Cl^-_{aq} . \qquad (3)$$

At equilibrium, the activities (a) of the reacting species are
related as follows:

$$K_1 = \frac{[a_{MCl_5{}^{2-}}]_{aq}}{[a_{M^{3+}}]_{aq}\, [a_{Cl^-}]^5_{aq}} , \qquad\qquad (2a)$$

$$K_2 = \frac{[a_{(R_4N)_2MCl_5}]_{org} \, [a_{Cl^-}]^2_{aq}}{[a_{R_4NCl}]^2_{org} \, [a_{MCl_5^{2-}}]^5_{aq}},$$ (3a)

where K_1 and K_2 are the equilibrium constants for the two reactions. The equilibrium distribution coefficient (K_d) for the trivalent metal, defined as the ratio of the activity sorbed on the resin to the activity in the aqueous phase at equilibrium, can be obtained by combination and rearrangement of equations (2a) and (3a):

$$K_d = \frac{[a_{(R_4N)_2MCl_5}]_{org}}{[a_{M^{3+}}]_{aq}} = K_1K_2 \, [a_{R_4NCl}]^2_{org} \, [a_{Cl^-}]^3_{aq}.$$ (4)

Thus, for the divalent anionic complex illustrated, the distribution coefficient, K_d, varies directly with the second power of the activity of the functional amine group (quaternary ammonium chloride) of the resin and with the third power of the aqueous chloride activity. Similar equations can be written to show that the K_d dependence on the activity of the functional amine group is first power for monovalent anionic complexes, second power for divalent complexes (as illustrated), third power for trivalent complexes, etc. However, the dependency on the activity of the aqueous chloride is third power for all complexes. The latter effect was confirmed experimentally by Hulet et al.[1]. Even though this is true, the dependency on the aqueous chloride concentration (rather than the activity) is much greater, since the activity coefficients increase rapidly with concentration in the region of interest. Baybarz and Weaver [2], in their studies of the Tramex system, found the K_d dependency to be proportional to the 17th power of the chloride concentration. Thus, the LiCl concentration in eluent solutions must be very carefully controlled to obtain the desired sorption and separations. Hulet et al. found that superior actinide-lanthanide group separations are obtained in the region of 10 M LiCl; the two series of elements tend to merge below 8 M LiCl, and the elution time becomes inconveniently long above 10 M. Their study also showed that, by increasing temperature from 25 to 87°C, resin cross-linkage to 8 or 10% divinylbenzene, and LiCl concentration to >10 M, the sorption and selectivity were improved; however, increasing the

HCl concentration above 0.1 $\underline{M}$, caused a significant decrease of sorption.

The comparative data obtained by Baybarz and Kinser ($\underline{6}$), in their study of the behavior of contaminant ions in Tramex extraction (from 11 $\underline{M}$ LiCl solutions) and stripping (with 0.5-10 $\underline{M}$ HCl) have been useful for planning and interpreting results from LiCl AIX operations. The distribution coefficients (K_d's) for the extraction of various ions of corrosion and fission product elements indicated that Ru^{3+}, Ce^{3+}, Eu^{3+}, Y^{3+}, Cr^{3+}, Ba^{2+}, and Sr^{2+} are sorbed more poorly than the tripositive actinides, that Zr^{4+}, MoO_4^{2-}, Ni^{2+}, and Pb^{2+} behave similarly to the actinides, and that Fe^{3+}, Co^{2+}, Mn^{2+}, Ti^{4+}, Cu^{2+}, Sn^{4+}, and Zn^{2+} are more strongly sorbed than the actinides. Stripping K_d's indicated that Ti^{4+} and Ni^{2+} are stripped with the actinides at any HCl concentration between 0.5 and 10 $\underline{M}$. However, most of the extracted elements can be left on the resin during the stripping of the actinides. The effects of various anions were also determined by Baybarz and Kinser. Increasing nitrate concentration was shown to cause an increase in extraction K_d's of both Am^{3+} and Eu^{3+}, but to cause a decrease in the separation factor between the two elements.

Process Equipment

Most of the equipment used at TRU for separating multigram quantities of actinides and lanthanides is contained on a compactly arranged equipment rack that is about 1 m wide and 2 m high. The rack is located within a heavily shielded, but small, hot cell (about 8 m^3 of space) which is equipped with a viewing window and a pair of master-slave manipulators. A schematic diagram of the equipment is shown in Fig. 1. The feed adjustment evaporator (25-L capacity) and the waste collection tank (70-L capacity) are located in a remote tank pit, and the eluent and resin addition tanks are located above the hot cell in a non-radioactive area. All of the equipment and piping which serve the feed and product solutions are built of glass or tantalum; these materials have provided excellent resistance to the highly corrosive, radioactive chloride solutions.

The calibrated, glass feed tank has a diameter of 102 mm and a capacity of 4.5 L. Solutions can be added to the tank via two routes: first, vacuum can be applied to the tank to motivate transfer of adjusted feed solution from the feed adjustment evaporator; and secondly, pressurized transfer of solution can be made from the eluent addition tank. A tantalum diaphragm pump is used to transfer solutions from the feed tank to the top of the ion exchange column. Flow rates are controlled by cycling vacuum and pressure against the diaphragm at a selected frequency; a discharge pressure of up to 200 kPa can be achieved.

The tantalum ion exchange column has a 50-mm diameter and is 76 cm long; it is heated to 70-80°C by hot water which is

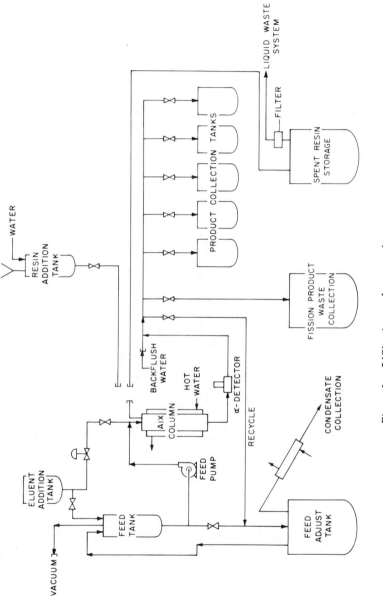

Figure 1. LiCl anion exchange equipment

circulated through the column jacket. Resin supports inside the
column are sintered tantalum discs. In the glass columns used
previously, a spring-loaded plate was provided at the top of the
column to prevent movement of the resin bed by rising bubbles of
radiolytically generated gas; however, the plate was found to be
unnecessary and was not provided in the present column.

Effluent solution from the ion exchange column is passed
through an in-line alpha-detector which activates a count-rate
meter and recorder outside the cell. Inside the detector, the
liquid flow is passed adjacent to a Mylar film-covered window,
which separates the liquid from the silicon diode detector.

In addition to the large waste collection tank, the column
effluent liquid can be routed to either of five product collection tanks, each of which has a capacity of 7 L.

Operating Procedure

Resin Preparation and Loading. Batches of 200-400 mesh
Dowex 1-X10 resin (chloride form) are classified to obtain a
middle fraction having a wet particle size range of about 55-
105 μm. The classified resin is treated with 6 $\underline{M}$ HCl to ensure
that the chloride form has been maintained. A 1.3-L volume of
the treated resin (measured after settling for 60 min) is
slurried in water and transferred into the ion exchange column.
The feed tank, pump, piping, column, and resin bed are conditioned by transfer of three column volumes of synthetic feed
solution (12 $\underline{M}$ LiCl--0.1 $\underline{M}$ HCl) through the system. This avoids
dilution of the actual feed solution by any solution left in the
equipment during previous operations.

Feed Pretreatment. A two-step clarification process is
used to eliminate problems with solids formation and to significantly reduce the amount of transplutonium elements diverted into
rework solutions by inclusion with the solid material (7). In
the first step, the feed solution (the product of a solvent
extraction process) is washed with an organic diluent (diethylbenzene) to remove entrained organic extractant that would be
degraded in the subsequent evaporation steps to a tar, which
could sorb a significant amount of the transplutonium elements.
The second step is designed to remove insoluble amounts of
aluminum, sodium, and zirconium, which are typical impurities
contained in the feed solutions. The treatment includes (i) ad-
justment to a small volume of 12 $\underline{M}$ LiCl--1 $\underline{M}$ HCl, (ii) filtration
to remove the insoluble materials, (iii) dilution by flushing the
filter with 12 $\underline{M}$ LiCl--1 $\underline{M}$ HCl solution, and (iv) readjustment to
a larger volume of 12 $\underline{M}$ LiCl--0.1 $\underline{M}$ HCl. This treatment is based
on the fact that the solubility of $AlCl_3$ is significantly lower
in 12 $\underline{M}$ LiCl--1 $\underline{M}$ HCl than in 12 $\underline{M}$ LiCl--0.1 $\underline{M}$ HCl.

Feed Adjustment. The feed solutions are adjusted to con-
centrations of 12.0 $\underline{M}$ LiCl and approximately 0.1 $\underline{M}$ HCl. The
typical feed volume used for a single loading and elution of the
resin is 3 L. Thus, when the loading is limited by total mass of
the actinides (to 35 g), the concentration is about 12 g/L; when
the limit is alpha-decay heat (54 W from 19 g of ^{244}Cm), the
power density is 18 W/L.

The specification for feed solution acidity is based on providing an HCl concentration high enough to prevent hydrolysis of
the actinides and yet low enough to allow sorption of the actinides within a narrow band at the top of the resin bed. By using
the maximum practical concentration of LiCl (12 $\underline{M}$), the highest
allowable concentration of HCl can be used. This is necessary
because the acidity is continuously reduced by radiolytic destruction of the HCl at such a rapid rate (typically about 0.01 $\underline{M}$/h)
that the technique of acidity adjustment, sampling, analysis, and
readjustment cannot be used. Thus, the acidity cannot be firmly
specified. The acceptability of the feed is achieved by using a
proven procedure for adjustment and is judged by observation of
the presence or absence of a precipitate.

The feed adjustment procedure includes (i) adding the
required amount of LiCl (if necessary), (ii) evaporating the
solution until a temperature of 142.5°C is reached (this is the
boiling point of 13 $\underline{M}$ LiCl), (iii) cooling to 50°C and adding
enough 12 $\underline{M}$ HCl to bring the feed solution acidity to 1 $\underline{M}$,
(iv) digesting the solution at 120 $\pm$ 0.5°C for 10 min (to dissolve hydrolyzed zirconium, which is frequently present, and to
lower the acidity to about 0.1 $\underline{M}$ HCl), and (v) cooling to 50°C.
The adjusted feed solution will remain stable for several hours,
which is enough time to complete the transfer to the ion exchange
column.

Loading. The appropriate volume (usually 3 L) of adjusted
feed is transferred to the feed tank and pumped to the ion
exchange column at a flow rate of 2 L/h (equivalent to a superficial velocity of 280 μm/s in the column). Elution of the resin
is begun immediately after the loading has been completed. A
movable detector is used to measure the neutron profile in the
column after the loading and at 1-hr intervals during the
elution. The position of the neutron peak (from ^{252}Cf) after
loading, as illustrated in Fig. 2, can indicate resin damage or
shrinkage at the top of the bed; a peak that is lower in the bed
is usually observed when the resin is used for more than one
loading and elution.

Elution. In general, the lanthanides are eluted first,
using four column volumes of 10 $\underline{M}$ LiCl; then, 90-95% of the americium and curium is eluted, using three column volumes of 9 $\underline{M}$ LiCl.
The remainder of the americium and curium is eluted along with all
of the heavier actinides, using two column volumes of 8 $\underline{M}$ HCl.

Finally, the column is flushed with three volumes of 0.8 $\underline{M}$ HCl to remove sorbed impurities, such as zirconium and plutonium. More precisely, the LiCl eluent solutions are adjusted to within 0.05 $\underline{M}$ of the desired concentration and are acidified to 0.10 $\pm$ 0.05 $\underline{M}$ HCl; hydroxylamine hydrochloride is added (to a concentration of 0.1 $\underline{M}$) as a reducing agent for any tetravalent cerium sorbed on the resin and methyl alcohol is added (to a concentration of 2.5% by volume) to suppress the rate of radiolytic destruction of HCl and the corresponding generation of radiolytic gases (8). Typically, the eluents are added to the feed tank in small portions and pumped to the column; thus, the eluents also serve to flush the feed system. The eluent flow rate is 1 L/h and the superficial column velocity is 140 μm/s.

The progress of the actinide elutions is monitored by periodic measurement of the neutron profile of the column, as illustrated in Fig. 2, and by a continuous reading of the alpha activity in the effluent solution from the column. A typical plot of the alpha activity is shown in Fig. 3. These data are used to select the routing of the effluent solution into the proper collection tank. Although a discriminator is used, the alpha-detector is usually sensitive to some of the beta-gamma emissions from the lanthanide fission products. Practically, the detector response may vary because of radiation exposure, heat effects, or physical damage. Also, the process separations vary from run to run because of the extreme sensitivity to solution concentrations and because of accumulated radiation damage to the resin. Thus, the operating procedure requires that the effluent solution be routed into a "pre-curium" collection tank when the alpha-detector reading rises above the background. If the reading continues to increase, the flow is routed into the curium product tank. If the alpha-detector malfunctions and no response occurs, route selections are based on solution volumes.

The location of berkelium, a beta emitter which cannot be monitored, can be estimated from the position of the californium in the column, as determined by the neutron peak (from ^{252}Cf), during the time that curium is in the effluent solution. Typical neutron peaks are shown in Fig. 2. By comparison of the relative distribution coeffients of the actinides, the berkelium location is known to be about midway between californium and curium. The last 5-10% of the americium-curium is purposely routed into the transcurium element product tank to minimize the berkelium loss. Subsequently, this americium-curium is recovered in a second-cycle LiCl AIX run.

Resin Damage, Replacement, and Storage. At the actinide loading limits described above, up to three loadings and elutions can be made on each batch of resin before accumulated resin damage becomes sufficiently extensive to prevent adequate separations. The exposure of the resin after three loadings and elutions of 19 g of ^{244}Cm (a total of 57 g of ^{244}Cm or 162 W of

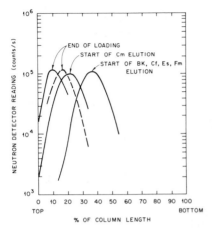

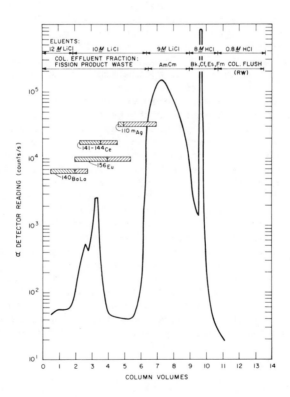

Figure 2. Neutron profiles of AIX column: (———) first loading and elution, (— —) second loading

Figure 3. Column effluent composition

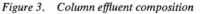

decay heat) has been estimated to be 2.5 MGy (2.5 x 10^8 rads) at
the top 15% of the resin bed and 1.5 MGy in the lower bed. This
is in the range of exposure that typically causes considerable
degradation to most anion exchange resins (9). Thus, in production operations at TRU, the resin must be replaced frequently.

Spent resin is transferred from the column through a 10-mm-
diam tube to a resin storage tank. The flush water continuously
overflows the storage tank through a screen to the ORNL Intermediate Level Liquid Waste System.

Process Use in Production Campaigns

The LiCl AIX process was used for several years at TRU to
treat product solutions from the continuous Tramex process. The
purpose of this treatment was to partition the americium and
curium from the heavier elements and to provide additional decontamination from fission products. However, the continuous Tramex
process was often plagued by malfunctions caused by feed solution
instability and emulsion formation in the pulsed columns. During
one campaign, a severe emulsion problem occurred and could not be
resolved. As an expedient, a Tramex batch extraction was
employed. The decontamination obtained from the lanthanide
fission products was about 5-10 times lower than that usually
obtained from the continuous multistage process. However, the
remaining lanthanides were efficiently removed by the subsequent
LiCl AIX process. These experiences showed that the combination
of a Tramex batch extraction with the LiCl AIX process provided
adequate decontamination of the transplutonium elements and
required less time and effort than the previous method. Thus,
the continuous version of the Tramex process was abandoned.

The process sequence now used is shown in Fig. 4. Since
only about 5% of the fission products are disposed of in waste
solutions from the Tramex batch extraction, that process serves
primarily as a feed pretreatment for the LiCl AIX. The Tramex
product contains about 98% of the transcurium elements and can be
processed quickly to maximize the recovery of ^{253}Es, which has a
20-d half-life. As time permits, the "clean rework" can be processed to recover the remaining actinides.

The sequence of process steps shown in Fig. 4 produces three
batches of spent resin (a total of about 4 L) and introduces
about 50 kg of LiCl into the liquid waste system. The initial
product recovery steps require about two weeks of operating time,
and the clean rework recovery requires another 2 weeks.

Process Results

A typical composition of feed solution and the fractional
distribution of the feed solution components into the various
exit streams are shown in Table I. The feed solution is usually
the product of a Cleanex batch solvent extraction (10), a process

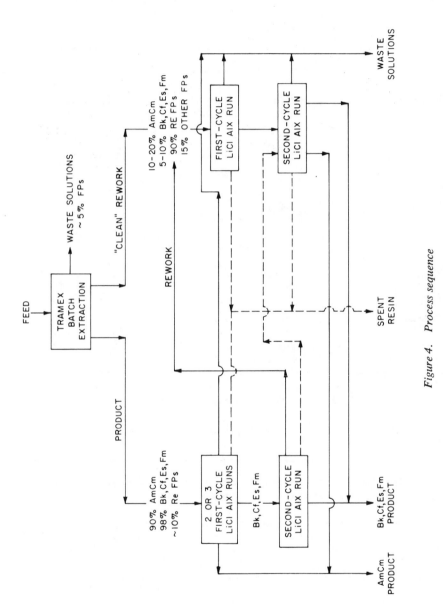

Figure 4. Process sequence

TABLE I. TYPICAL COMPOSITION OF FEED SOLUTIONS AND DISTRIBUTION
          OF COMPONENTS TO EXIT SOLUTIONS

Component	Amounts in Feed Solution	% Distribution to Exit Solutions			
		Products		Rework[a]	F.P. Waste
		AmCm	BkCfEsFm		
Part A: Transplutonium Elements					
^{243}Am	0.1 g	98	0.3[b]	1.5	0.2
^{244}Cm (Total Cm)	20(50)g	98	0.3[b]	1.5	0.2
^{249}Bk	35 mg	0.5	98	1.5	<0.1
^{252}Cf	400 mg	0.05	99	1.0	<0.1
^{253}Es	2.5 mg	<0.1	99	1.0	<0.1
^{257}Fm	0.8 pg[c]				
Part B: Fission Products					
^{95}ZrNb	0.2 TBq	<1	(d)	<1	100[e]
^{103,106}Ru	3 TBq	<0.1	(d)	<1	100[e]
^{110m}Ag	0.4 TBq	0.3	(d)	<1	100
^{140}BaLa	10 TBq	0.2	(d)	<1	100
^{141,144}Ce	70 TBq	0.03	(d)	10	90
^{156}Eu	40 TBq	0.02	(d)	10	90
Gross Gamma	1×10^{15} c/m	0.03	(d)	10	90
Part C. Macroscopic Impurities					
Al	30 g	<1	<1	70	30
Na	60 g	<1	<1	90	10

a
  Short-lived nuclides are lost by radioactive decay during
   typical 0.5-year storage of rework solution.
b
  Recovered by subsequent process step.
c
  Estimated.
d
  Cannot be measured because of masking by ^{252}Cf spontaneous
   fission spectrum.
e
  Removed in Tramex solvent.

in which most of the nonlanthanide fission products and macroscopic corrosion products are removed. However, the Cleanex process does not effectively remove residual aluminum from the target rod dissolution and introduces a significant amount of sodium salts into the product solution. (NaOH is used to neutralize excess acidity in the Cleanex process.) The aluminum and sodium impurities are only slightly soluble in concentrated LiCl solutions; however, the feed pretreatment filtration step, described above, has removed these insoluble impurities effectively.

In some production campaigns, the "clean" rework solutions, as well as the "dirty" rework, have been stored for about six months and then processed along with the subsequent group of irradiated targets. However, during the storage period, small but significant amounts of the short-lived products, especially ^{253}Es and ^{257}Fm but also including some of the ^{249}Bk, have been lost by radioactive decay. Thus, the practice of reprocessing the "clean" rework solution immediately after the initial actinide recovery steps has increased product yields.

Conclusions

The LiCl AIX process has been successfully adapted to the multigram scale and has been used effectively in transuranium-element production campaigns to separate the lanthanide fission products from the transplutonium actinides and to partition americium and curium from the heavier elements. Corrosion of the tantalum and glass equipment has been negligible. Although radiolytic gas generation has not caused a problem, radiation exposure of the Dowex 1-X10 anion exchange resin does occur significantly. However, the 1.3-L resin bed can be used successfully to process up to three batches, each containing 19 g of ^{244}Cm (54 W of decay heat). The chromatographic elution process is controlled by use of an alpha detector in the column effluent line and by periodic measurement of the neutron profile of the column. The development and use of feed pretreatment and operating methods have enabled effective and dependable operation.

Acknowledgement

This research was sponsored by the Office of Basic Energy Sciences, U.S. Department of Energy, under contract W-7405-eng-26 with the Union Carbide Corporation.

Literature Cited

1. Hulet, E. K.; Gutmacher, R. G.; and Coops, M. S. "Group Separation of the Actinides from the Lanthanides by Anion Exchange," *J. Inorg. Nucl Chem.*, 1961, *17*, 350-360.

2. Baybarz, R. D.; Weaver, B. "Separation of Transplutoniums
    from Lanthanides by Tertiary Amine Extraction," ORNL-3185
    (December 1961).
3. Leuze, R. E.; Baybarz, R. D.; Weaver, B. "Application of
    Amine and Phosphonate Extractants to Transplutonium Element
    Production," Nucl. Sci. Eng. 1963, 17, 252-258.
4. Weaver, B. J. Inorg. Nucl. Chem., 1968, 30, 2223.
5. Baybarz, R. D.; Orr, P. B. "Final Purification of the
    Heavy Actinides from the First Four Campaigns of the TRU
    Program," ORNL-TM-2083 (November 1967).
6. Baybarz, R. D.; Kinser, H. B. "Separation of Transplutoniums
    and Lanthanides by Tertiary Amine Extraction II. Contami-
    nant Ions," ORNL-3244 (February 1962).
7. Collins, E. D.; Bigelow, J. E. "Chemical Process Engineering
    in the Transuranium Processing Plant," Proc. 24th Conf.
    Remote Syst. Technol., 1976, 130-139.
8. Baybarz, R. D. "Alpha Radiation Effects on Concentrated LiCl
    Solutions Containing HCl, and the use of Methanol as an
    Inhibitor of Acid Radiolysis," J. Inorg. Nucl. Chem., 1965,
    27, 725-730.
9. Gangwer, T. E.; Goldstein, M.; Pillay, K. K. S. "Radiation
    Effects on Ion Exchange Materials," BNL-50781, Brookhaven
    National Laboratory (November 1977).
10. Bigelow, J. E.; Collins, E. D.; King, L. J. "The "Cleanex"
    Process: A Versatile Solvent Extraction Process for
    Recovery and Purification of Lanthanides, Americium, and
    Curium," Actinide Separations, ACS Symp. Series, No. 117,
    1980, 147-155.

RECEIVED December 19, 1980.

Chromatographic Cation Exchange Separation of Decigram Quantities of Californium and Other Transplutonium Elements

D. E. BENKER, F. R. CHATTIN, E. D. COLLINS, J. B. KNAUER,
P. B. ORR, R. G. GOSS, and J. T. WIGGINS

Oak Ridge National Laboratory, Oak Ridge, TN 37830

High-pressure cation exchange is used routinely in the
Transuranium Processing Plant (TRU) at Oak Ridge National
Laboratory (ORNL) to separate decigram quantities of transpluto-
nium elements that have been produced in the High Flux Isotope
Reactor. The process is based on chromatographic elution from
Dowex 50W-X8 resin using ammonium alpha-hydroxyisobutyrate
(AHIB) as the eluent (1). Since 1967, a total of 4.6 g of
^{252}Cf, 0.5 g of ^{249}Bk, 19 mg of ^{253}Es, and 10 pg of ^{257}Fm
(estimated) has been separated at TRU using this procedure.
This process, which was originally developed at ORNL by Campbell
and Buxton (2,3), was later adapted for processing at high activity levels at TRU by Baybarz et al. (4). High pressure is
required so that very fine resin and high flow rates may be
used. The fine resin is necessary to provide the resolution for
good separation of the actinides, while the high flow rate is
needed to mitigate radiation damage. Also, the use of high
pressure suppresses radiolytic gassing.

This paper discusses the equipment and process steps used
in the initial separation of transplutonium elements and presents some typical results.

Process Description

The transplutonium elements are separated at TRU using
elution development chromatography, which is normally the preferred type of process for separating small quantities of
these elements (5). In the procedure used at TRU, the trivalent
actinides are first sorbed from a dilute (<0.5 M) HNO_3 solution
onto cation resin. The resin is then converted to the ammonium
form, and the actinides are then eluted from the resin with a
complexing agent. The resolution of the actinides into individual bands as they are eluted through the ion exchange column is
accomplished by the difference in the complexing strength of the
actinides with the complexing agent--the organic anion provided
by the eluent, AHIB. The elution position of the elements (or

0097-6156/81/0161-0161$05.00/0

the speed with which the elements are eluted through the column)
is strongly dependent on the concentration of the complexing
anion. In order to obtain consistent results with this process,
the concentration of the complexing anion must be strictly regu-
lated by careful control of the AHIB molarity and pH of the
eluent solutions. This sensitivity of the actinide separation
to the pH of the eluents is the reason why the resin is converted to the ammonium form before an elution.

Equipment

The ion exchange equipment, which is located on a metal
rack (1.8 x 0.9 x 0.4 m) in one of the TRU hot cells, is maintained and operated with two master-slave manipulators that
penetrate the 1.4-m-thick shielding wall. A flow diagram is
shown in Fig. 1.

Two columns are used. The short column, 14 mm in diameter
and 0.2 m high, is used for the initial loading of the feed
solution which is transferred from a vacuum-pressure pot
(about 1-L capacity). This column contains 35 mL of resin and
is not heated.

The long column, 13.4 mm in diameter by 1.2 m high, is
used for the chromatographic separation of the transcurium
elements. After the short column has been loaded, the two
columns are connected, and eluent solutions are pumped through
the short column to the long column. A positive-displacement
pump (Beckman Co. Accuflo No. 312880), powered by an in-cell
electric motor, is used to transfer the eluent solutions
through the columns. The long column contains 145 mL of resin
and is jacketed for heating to about 70°C with hot water.
Hot water for equipment in the hot cells is usually recirculated
through heaters and pumps that are located in a nonradioactive
operating area. However, in this case, the operating pressure
of the column (up to 5 MPa) is higher than the water pressure in
the jacket (0.4 MPa) and, if a leak were to develop in the
column wall, radioactive process solution would be forced into
the jacket and thence into the nonradioactive operating area.
In order to protect personnel from a potentially serious
radiation hazard, the water from the jacket, is used on a once-
through basis and then discharged to a tank used for collecting
radioactive waste solutions. This practice is followed even
though the water is not contaminated under normal operating
conditions.

The ion exchange columns, the vacuum-pressure pot, and the
piping and valves are all made of 304 (or 304L) stainless steel.
The resin bed in each column is supported by a 304 (or 304L)
stainless steel frit (5 μm) at the bottom of each column.

A flow-through alpha detector in the column effluent line
is used to monitor the progress of an elution, and a movable
neutron probe is used to determine the location of the ^{252}Cf

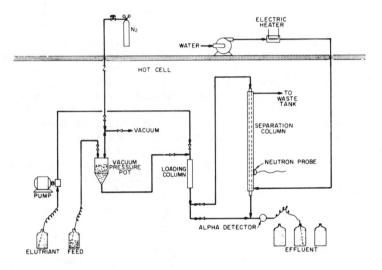

Figure 1. Diagram of pressurized ion exchange system used to separate transplutonium elements

band. (Other transplutonium isotopes in TRU solutions generate
neutrons, but the quantity is trivial when compared to the
neutrons from milligram quantities of ^{252}Cf.) Both detectors
are silicon surface-barrier diodes (supplied by Harshaw Chemical
Company, Crystal and Electronics Department, Division of
Kewannee Oil Company, Solon, Ohio, 44139; or by EG&G ORTEC, Oak
Ridge, Tennessee, 37830) (6), and each is wired to a count-rate
meter and a recorder outside the hot cell. The diode for the
neutron probe is covered with a thin piece of polyethylene and
measures protons from an n,p reaction.

Reagents

The cation exchange resin is Dowex 50W-X8 which has been
hydraulically classified. The chromatographic column is filled
with resin in the size ranges of 28-35, 35-42, or 42-56 μm,
while the loading column is filled with 56-70 μm resin. Before
beginning a separation run, the resin is converted to the ammonium form by forcing 0.3 $\underline{M}$ NH$_4$NO$_3$ through the resin beds.
The eluent is prepared by dissolving a weighed amount of
alpha-hydroxyisobutyric acid (manufactured by Eastern Chemical
Corp., Division of Guardian Chemical Corporation, 230 Marcus
Blvd., Hauppauge, New York, 11787) in demineralized water,
adjusting the pH with ammonium hydroxide, and then diluting with
demineralized water to the final volume. Eluent is ordinarily
prepared a couple of days before a series of runs, but solutions
have been stored for as long as four weeks in the hot cell
before being used.

Processing Steps

Feed Preparation. Feed solutions are adjusted to less than
2 L of ≤0.5 $\underline{M}$ HNO$_3$ before processing. Since the feed is the
product of a LiCl-based anion exchange run (7), the solution con-
tains both LiCl (about 3 $\underline{M}$) and HCl (about 1 $\underline{M}$). The lithium
and chloride ions are removed, and the solution is converted to
the nitrate form by precipitating the metal hydroxide with 1 $\underline{M}$
LiOH, filtering through a glass frit, washing with water, re-
dissolving with strong HNO$_3$ (8 and 16 $\underline{M}$), and finally diluting
with 0.1 $\underline{M}$ HNO$_3$ solution to make a 0.25 $\underline{M}$ HNO$_3$ solution. In
order to improve the actinide recovery during the filtration
step, up to 200 mg of ferric iron (as FeCl$_3$·6H$_2$O) is added as a
carrier before the precipitation. Also, a crystal growth period
of 90 min between the precipitation and the filtration steps
has been found to significantly improve the filtration step by
increasing the filtration rate and reducing the amount of actinides in the filtrate and left undissolved on the filter.

Feed Loading. The adjusted feed solution is transferred
by vacuum into the vacuum-pressure pot. Then, the pot is

pressurized to 0.3 MPa with N_2, and the feed solution is forced
through the loading column at 2.7 mm/s (16 mL/cm^2·min). The
elution column is bypassed. The actinides form a tight band at
the top of the column, which can be verified by locating the
^{252}Cf with the neutron probe. The resin is then washed with
water and 310 mL of 0.3 $\underline{M}$ NH_4NO_3 to displace H^+ ions that would
interfere with the actinide separation during the elution. Very
little movement of the actinide band takes place during the
washing step.

The use of the loading column is desirable for three
reasons. First, the pressure resistance in the short column is
considerably less, so it is not necessary to use the high-
pressure pump to transfer the highly radioactive solutions.
This facilitates eventual pump repairs in a glove-box facility.
Second, the most severely radiation-damaged resin is the relatively small volume contained in the short column. This resin
can be replaced after each loading, while the larger volume in
the long column can be used for several elutions. This mini-
mizes the volume of waste resin. Third, the acidic loading raffinate solution is not passed through the long column and, thus,
the relatively large volume of resin in that column remains in
the ammonium form necessary for elution. This lessens the
radiation damage to the resin by reducing the time required for
the NH_4NO_3 wash and allows larger batches of ^{252}Cf (39 W/g) to
be processed during each loading and elution. (For a typical
batch of feed, ^{252}Cf is the primary contributor to radiation
exposure.) By using the smaller loading column, feed batches
containing up to 200 mg of ^{252}Cf have been processed
successfully. The largest run that has been made contained 380
mg of ^{252}Cf. During that run, about 10% of the actinides did
not elute from the loading column because of severe radiation
damage to the resin, and the actinides had to be recovered with
a strong acid leach of the resin. Several runs have been made
with 240 to 320 mg of ^{252}Cf; however, even though all the californium was eluted, the elution bands were spread to the point
that the separations achieved were not entirely effective. The
intermediate fractions between the einsteinium and californium
peaks contained from 10 to 50% of the einsteinium and californium in the feed and had to be recycled.

Elution. After the actinides have been loaded and the resin
washed, the (short) loading column effluent is routed to the
(long) separation column, and the high-pressure pump is used to
transfer eluent through the columns at a rate of 1.0 L/h (a
superficial velocity of 2.0 mm/s through the elution column).
At this rate, the pressure drop through the two columns is 2 to
5 MPa. The eluent solutions used are: 220 mL of 0.25 $\underline{M}$ AHIB--
pH 3.9 (to elute the actinides from the loading column onto the
separation column), about 1.5 L of 0.25 $\underline{M}$ AHIB--pH 4.2 (to
elute all the fermium, einsteinium, and californium), 700 mL of

0.25 $\underline{M}$ AHIB--pH 4.6 (to elute berkelium), 450 mL of 0.50 $\underline{M}$
AHIB--pH 4.8 (to strip americium and curium), and finally, 450
mL of water (to flush reagents from the equipment).
The separation column effluent is divided into about 15
fractions that are collected in small (250-mL) polyethylene
bottles. The volume collected in each bottle is determined by
the appearance of the alpha-emitting elements in the column
effluent solution as indicated by the response from the flow-
through alpha detector; a typical response curve is shown in
Fig. 2. Normally, two einsteinium fractions, two intermediate
fractions, and three californium fractions are collected. The
intermediate fractions are taken when the valley between the
einsteinium and californium peaks occurs on the response curve
and usually contain less than 5% of each element. Sometimes the
alpha trace will show a small fermium peak just ahead of the
einsteinium, but usually there is not enough fermium alpha to
make a response and the fermium is assumed to be in one or both
of the two fractions taken just prior to the einsteinium. The
berkelium is primarily a beta emitter and is not detected by
either the alpha or neutron detectors; thus, three fractions are
usually taken after the californium alpha peak to isolate the
berkelium. If there is a significant amount of ^{244}Cm in the
feed (milligram quantities), the alpha trace will show a third
major peak when americium and curium are eluted at the end of
the run.
Monitoring of the ^{252}Cf band movement with the neutron
probe (Fig. 3) provides advance information on when to expect
einsteinium in the eluate and can be used as a backup indicator
for changing effluent collection bottles during the einsteinium-
californium elution if the alpha detector fails. Also, the
neutron probe can be used to detect problems such as insufficient resin in the columns or incomplete elution from the
loading column.

Results

The distribution of einsteinium, californium, berkelium,
and curium for a typical batch of feed is shown in Table I.
The einsteinium and berkelium product fractions are usually
decontaminated from ^{252}Cf by factors of 10^3-10^4 and 10^2-10^3,
respectively; and the californium product fractions are usually
decontaminated from ^{253}Es by a factor of 10^2. The maximum concentrations of ^{252}Cf, ^{249}Bk, and ^{253}Es in the respective product
fractions are typically about 0.4 g/L, about 50 mg/L, and about
3 mg/L, respectively.

Resin Damage and Disposal

Radiation damage to the resin is more severe in the loading
column. If the actinides load up to the resin capacity (about 2

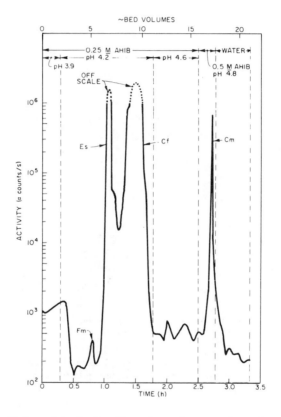

Figure 2. Alpha activity in the effluent from the separation column as measured by the flow-through alpha detector during a typical run

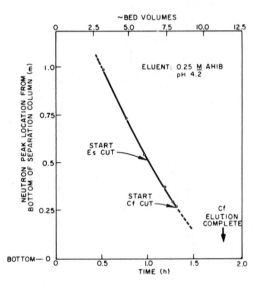

*Figure 3.   Movement of the ^{252}Cf peak as measured by the neutron probe during a
typical run to separate transplutonium elements*

Table I. Actinide Distribution for Typical Run

	^{253}Es	^{252}Cf	^{249}Bk	^{244}Cm
Feed, mg	0.6	150	15	25
% in Es product	96.3	0.05	--	--
% in Cf product	1.5	98.2	0.3	--
% in Bk product	0.1	0.3	99.5	--
% in rework	2.1	1.5	0.2	100

eq/L), the radiation dose for a typical run would be about 20 MGy (2×10^9 rads). Because of this high radiation dose, the resin in the loading column is replaced after each run. The resin is discharged by removing the bottom fitting from the column and slurrying the resin out with water. The appearance of the used resin is similar to fresh resin except that the top 10% of the resin bed is black. Resin from the elution column is removed in a similar manner, usually after a total of 300-400 mg of ^{252}Cf has been processed (two to three runs). The estimated total radiation dose to this resin is 0.8 MGy (8×10^7 rads). However, during the most recent processing campaign, a total of 830 mg of ^{252}Cf and 3.3 mg of ^{253}Es was satisfactorily processed in five runs with one batch of resin; this is equivalent to an estimated dose of 2 MGy (2×10^8 rads). The appearance of the resin from the elution column is normally uniform in color and only slightly darker than that of fresh resin.

The used resin from each column may contain a small amount of actinides that failed to elute. The amount is estimated by counting fast neutrons from ^{252}Cf with the in-cell neutron probe. The ^{252}Cf content of the resin, usually <100 μg of ^{252}Cf, can be recovered by leaching with strong HNO_3. After acid leaching several batches of resin have been calcined in an electric resistance furnace and the ashes leached with HNO_3; however, very little additional ^{252}Cf was recovered.

Additional Separations

The einsteinium-fermium and californium fractions collected during the initial separation described above are processed by means of a second cycle of high-pressure chromatographic cation exchange for additional partitioning of the actinide elements. The equipment and processing steps are similar to those described above.

The einsteinium-fermium fractions are taken to a shielded cave (a small hot cell), where the second-cycle run is made to remove residual californium (10-200 µg of ^{252}Cf). The elution column in the cave is 8 mm in diameter and 1.2 m high and contains 21 mL of resin. Each loading and elution is usually limited to <500 µg of ^{253}Es; the operating time required is 4 to 6 h. One run is usually sufficient to reduce the ^{252}Cf content in the einsteinium-fermium fraction to <1 µg. This represents, for the combined initial and second-cycle run, an overall decontamination factor of californium from einsteinium of about 10^5. With a ^{252}Cf content of <1 µg, the einsteinium-fermium can be taken to a glove box for further purification before being packaged for shipment.

The californium fractions are taken to another hot cell [the TURF Californium Facility (8)], where second-cycle runs are made to further decontaminate the californium from ^{244}Cm and to recover a second batch of ^{253}Es, which grows into the californium from the beta decay of ^{253}Cf (18-d half-life). These runs are limited to <125 mg of ^{252}Cf and require 5 h of operating time. The purified californium is stored for about 2 years and then reprocessed by means of another high-pressure ion exchange run to recover ^{248}Cm, the alpha-decay daughter of ^{252}Cf (2.6-y half-life). The ^{248}Cm is more desirable to some researchers than the more readily available ^{244}Cm because of its very low specific activity (0.2 GBq/g for ^{248}Cm as compared with 3 TBq/g for ^{244}Cm). Since 1970, about 0.8 g of ^{248}Cm has been recovered.

Summary

Decigram quantities of highly radioactive transplutonium elements are routinely partitioned at TRU by chromatographic elution from cation resin using AHIB eluents. Batch runs containing up to 200 mg of ^{252}Cf can be made in about 5 h (2 h to load the feed and 3 h for the elution), with two high-pressure ion exchange columns, a small one for the initial loading of the feed and a large one for the elution. The separations achieved in the column are preserved by routing the column effluent through an alpha detector and using the response from the detector to select appropriate product fractions. The high-pressure ion exchange process has been reliable and relatively easy to operate; therefore it will continue to be used for partitioning transplutonium elements at TRU.

Acknowledgment

This work was sponsored by the Office of Basic Energy Sciences, U. S. Department of Energy, under contract W-7405-eng-26 with the Union Carbide Corporation.

Literature Cited

1. Choppin, G. R.; Harvey, B. G.; Thompson, S. G. J. Inorg. Nucl. Chem., 1956, 2, 66.
2. Campbell, D. O.; Buxton, S. R. Ind. Eng. Chem., Process Des. Dev., 1970, 9, 89.
3. Campbell, D. O. Ind. Eng. Chem., Process Des. Dev., 1970, 9, 95.
4. Baybarz, R. D.; Knauer, J. B.; Orr, P. B. "Final Isolation and Purification of the Transplutonium Elements from the Twelve Campaigns Conducted at TRU During the Period August-1967 December 1971," ORNL-4672 (1973).
5. Campbell, D. O., "The Application of Pressurized Ion Exchange to Separations of Transplutonium Elements," Industrial-Scale Production-Recovery-Separation of Transplutonium Elements Symposium, ACS Second Chemical Congress, Las Vegas, Nevada, 1980.
6. Zedler, R. E. Oak Ridge National Laboratory, Personal Communication, 1980.
7. Collins, E. D.; Bigelow, J. E. "Chemical Process Engineering in the Transuranium Processing Plant," Am. Nucl. Soc., Proc. 24th Conf. Remote Syst. Technol., 1976, 130.
8. Peishel, F. L.; Burch, W. D.; Jarvis, J. P. "Design and Installation of the TURF Californium Facility," Am. Nucl. Soc. Proc. 18th Conf. Remote Syst. Technol., 1970, 93.

RECEIVED December 19, 1980.

Preparation of Curium–Americium Oxide Microspheres by Resin-Bead Loading

F. R. CHATTIN, D. E. BENKER, M. H. LLOYD, P. B. ORR, R. G. ROSS, and J. T. WIGGINS

Oak Ridge National Laboratory, Oak Ridge, TN 37830

A resin-bead loading and calcination technique developed at the Transuranium Processing Plant (TRU) of the Oak Ridge National Laboratory (ORNL) is used routinely for producing uniform particles of curium-americium oxide in the size range desired for the fabrication of targets for irradiation in the High Flux Isotope Reactor (HFIR). TRU is the storage, production, and distribution center for the heavy-element research program of the U. S. Department of Energy. Target rods are remotely fabricated at TRU, irradiated in the HFIR, and then processed chemically at TRU for the separation and purification of the heavy actinide elements. Berkelium, californium, einsteinium, and fermium are distributed to researchers. Curium and americium that are recovered during the chemical processing are refabricated into targets for additional irradiation. About 200 g of curium-americium oxide are produced annually for use in fabricating HFIR targets. Detailed descriptions of TRU and the overall production program have been published previously (1,2, 3,4,5), and a summary of the program is presented in another paper at this symposium (6).

A HFIR target is a 9.4-mm-diam by 0.89-m-long aluminum rod encased in a cylindrical aluminum tube which serves to channel the flow of cooling water in the reactor. The central target rod contains some end hardware plus thirty-five 6.3-mm-diam by 14-mm-long pellets that contain a blend of curium-americium oxide (~15% by volume) and aluminum powder encased in an aluminum jacket. A pellet is made by placing a weighed amount of blended actinide oxide--aluminum powder in a liner of aluminum tubing that has had one end closed with aluminum powder pressed to theoretical density. Then clean aluminum powder is added to form a top cap, and the pellet is cold-pressed in a pellet die.

The size of the curium-americium oxide particles is an important criterion in the production of HFIR targets because of the relatively high oxide content in the blend. The oxide particles must be uniformly dispersed in the aluminum pellets, and the pressed pellets must have a continuous aluminum phase to

0097-6156/81/0161-0173$05.00/0

ensure adequate heat transfer during irradiation. When oxide
particles at some size smaller than 10 μm in diameter are mixed
with -325 mesh (<44 μm) aluminum powder and pressed into
pellets, the oxide phase may be continuous and the thermal conductivity low. With oxide particles ranging from 20 to 210 μm
in diameter, the aluminum phase is continuous and the thermal
conductivity is satisfactory for irradiation in high neutron
fluxes. The resin-bead loading and calcination technique allows
this criterion to be met easily even though remote operation is
necessary for curium oxide production.

In the following sections, a brief chronology of the development of this process is presented; the materials, equipment,
and basic operations relating to the resin-bead loading and
calcination method of producing sized curium-americium oxide
microspheres at TRU are described; and typical production data
are presented.

Process Development

During the mid-1960s, the feed material for the HFIR
targets was ^{242}Pu. Plutonium oxide was prepared in glove-box
operations using hydroxide precipitation, calcination, and
grinding (7). Process development studies made with rare earths
suggested that curium oxide prepared by hydroxide or oxalate
precipitation methods would not be suitable for HFIR targets.
Therefore, a sol-gel method for preparing curium oxide remotely
was developed and was used from 1968 through 1970 (8). However,
the sol-gel process was not well-suited for the small batch
sizes that are involved in the TRU program. Equipment startup
and shutdown were a major part of the operating time. The processing was plagued with erratic operation, poor yields, and
production of oversize oxide particles that required undesirable
grinding and screening operations. A simpler, more reliable
oxide preparation method that was adaptable to remote operation
was needed.

The resin-bead loading and calcination process was developed and evaluated during 1971 in a series of test runs in
which a total of 150 g of curium oxide was successfully prepared
and subsequently used in HFIR targets. Two resins, which required different process steps, were tested. The process using
Dowex 50W resin, which contains sulfonic acid exchange groups,
was selected for continued curium oxide production over a similar process using Amberlite IRC-50 resin, which contains car-
boxylic acid exchange groups. The Dowex 50W process (Fig. 1)
requires fewer steps and is better suited for remote operation.
In the Dowex 50W process, pre-sized Dowex 50W-X8 resin beads are
loaded to saturation from a dilute HNO_3 solution of curium-
americium. The resin is rinsed with water, and then the resin
matrix is destroyed by calcination to form dense actinide oxide
microspheres. Curium-americium oxide for HFIR targets has been

produced exclusively by this process since 1971. Until 1975,
the production was accomplished in special production runs in
the development equipment. In 1975, the present production-
scale equipment was installed in the hot cells.

Materials

Feed Solutions. The process is applicable to both americium and curium, which are not usually separated from each other
at TRU. The curium and americium recovered from the processing
of irradiated HFIR targets are supplemented with additional
americium or curium obtained from other programs. The supple-
mental materials may be blended with the recycle material or
processed separately. The curium and/or americium may be processed through a variety of purification steps, depending on the
source of the material, to remove other actinides, rare earths,
or ^{240}Pu, the decay daughter of ^{244}Cm. However, the final step
before oxide production is always a double oxalate precipitation
for removal of ionic contaminants, and the product from that
step is collected in strong HNO_3 in a tantalum-lined evaporator.
All curium-americium solutions for a series of oxide production
runs (typically containing 50-100 g of actinides) are collected,
mixed, and sampled to determine the americium and curium isotopic compositions of the feed and ultimately the isotopic composition of the actinides in HFIR targets prepared from the
oxide products.

Resin Beads. The resin beads are prepared from commer-
cially available Dowex 50W-X8, a sulfonic acid cation exchange
resin. The 8% cross-linked resin was chosen because it does not
shrink or expand excessively in acid-water systems and because
it is readily available. Resin cross-linkage had no detectable
effect on the final oxide products in the development test runs.
The resin is hydraulically classified to produce a size fraction
that is predominately 60-80 μm in diameter. The classified
resin is washed with 6 M HCl to remove contaminants such as iron
and rinsed thoroughly with water.

Equipment

The equipment for curium-americium oxide production (illustrated schematically in Fig. 2) is located on a rack within a
master-slave manipulator-equipped hot cell in the TRU cell
bank. The feed adjustment and raffinate collection vessels are
located in a tank pit in another part of the cell bank. General
design considerations and operating philosophies for chemical
process operations at TRU have been described previously (4).
Only equipment items that are unique to the curium-americium
oxide production are discussed below.

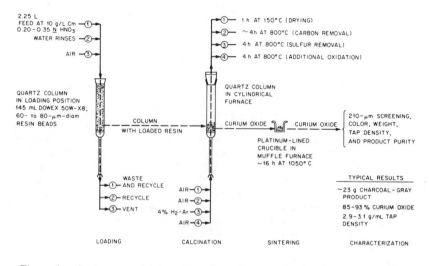

Figure 1. Curium–americium oxide microsphere production by Dowex 50W-X8 resin-bead loading and calcination

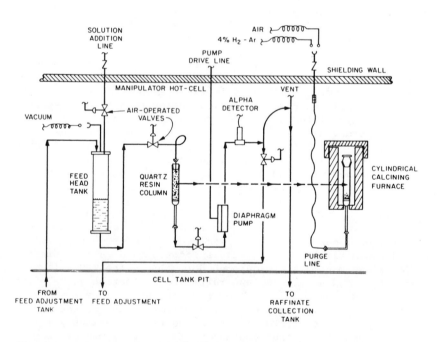

Figure 2. Schematic arrangement of equipment for the production of curium–americium oxide microspheres

Resin Loading Equipment. The feed is prepared in a 25-L tantalum-lined evaporator. Small, run-size batches (about 22.5 g of actinides) of adjusted feed solution are transferred by vacuum to a 4-L glass-pipe head tank on the equipment rack. The feed solution is then drawn through a quartz column containing the resin beads. A diaphragm pump is used to control the rate for loading the actinides on the resin. The effluent solution from the column passes through the pump and is discharged through an in-line alpha detector which is used to determine when curium breakthrough occurs. The effluent solution is collected in a 25-L Zircaloy-2 raffinate tank until breakthrough occurs and is diverted back into the tantalum-lined feed adjustment tank following breakthrough. In this way, actinides in the column effluent can be conveniently recycled to the feed for the subsequent run. This equipment arrangement facilitates complete loading of the resin without loss of curium from the oxide production effort. This was not possible in the development equipment, and about 5% of the resin usually was not loaded.

The overall length of the quartz column is 0.4 m. The upper portion of the column has an inside diameter of 30 mm and is 0.25 m long. About 145 mL of resin can be supported on the quartz frit at the bottom of the upper section. The 0.15-m-long by 12-mm-diam lower portion is provided for access to the bottom of the quartz column when it is placed in the cylindrical furnace for the calcination of the resin. Both ends of the column have ground ball joints for connections to other equipment. For loading operations the column is clamped to the equipment rack and is connected with flexible polyethylene tubing. The top of the column can either be connected to the feed head tank or be left open for adding resin and solutions directly to the column. Each quartz column is normally reused throughout a complete series of consecutive runs, usually five to nine runs, before it is discarded.

Calcination and Product Characterization Equipment. The resin drying, burning, sulfur removal, and oxidation steps are carried out in a cylindrical furnace built specifically for heating the quartz column. A Kanthal heating element is used to provide adequate furnace lifetime. The 63-mm-diam by 0.3-m-high firing cavity has a removable top and a 25-mm-diam access hole in the bottom. In the firing position, the lower section of the column extends through the bottom of the furnace and a flexible polyethylene gas purge line is attached well away from the heat-affected region through a tempered glass tube extension. A quartz frit top is placed on the column to provide an exit path for the purge and combustion gases while preventing the resin from "popping out like popcorn" during the drying step.

The sintering step (to 1050°C) is carried out in a laboratory muffle furnace with the curium-americium oxide contained in an open 50-mL platinum-lined Inconel crucible.

The oxide characterization steps are carried out with laboratory equipment that has been adapted as necessary to facilitate
handling in-cell with the manipulators. The items include a
210-μm-opening stainless steel screen assembly, a 300-g-capacity
triple-beam balance, a 10-mL glass graduated cylinder and as-
sorted weigh pans, spatulas, and oxide containers. A remotely
operated calorimeter is used to assay the ^{244}Cm content.

Processing Steps

Feed Adjustment. The tantalum-lined evaporator used to
collect the actinide product solutions from a series of double
oxalate precipitation runs also serves to adjust the composite
product to a feed solution for a series of oxide production
runs. Excess acid is removed by boiling the solution slowly to
near dryness. The temperature is held at greater than 119°C for
5 h or more during boiling to ensure the destruction of all
oxalates. After approximate dryness has been reached, the evaporator is cooled, 0.01 $\underline{M}$ HNO$_3$ is added to dilute the actinides
to less than 10 g/L (usually to 10-15 L, total volume) and a
sample is taken to determine the acidity (typically 0.05-0.10 $\underline{M}$)
and to verify that the actinides are still in solution. The
adjustment is completed by the addition of acid and evaporation
to give an actinide concentration of about 10 g/L and an acid
concentration of 0.20-0.35 $\underline{M}$ at the final volume.

The acidity and actinide concentrations of the feed solution have to be readjusted between loading runs because the
breakthrough raffinates and rinse solutions are recycled to the
feed adjustment tank during the resin loading step and various
solutions are added during the preparations for the next run of
the series. When the quartz column is returned for the subsequent loading run, it contains a small amount (up to 200 mg)
of curium-americium oxide on the frit. To prevent this material
from being dissolved in the raffinate and lost, the column is
given a preliminary leach with 250 mL of 0.25 $\underline{M}$ HNO$_3$, which is
returned to the feed adjustment tank. Then, a new batch of
resin is poured directly into the column as a slurry. The
excess water is pumped to the feed adjustment tank. Finally,
the top connection to the column is made and 200 mL of water is
transferred through the system as the final test of the flow
path in preparation for the next loading run. An accounting is
kept of the acid and actinides recycled, and the readjustment is
made by evaporation between runs and dilution at the start of
each loading run without repeating the original feed adjustment
procedure and without resampling. The acid concentration is
held in the required range, but the curium concentration is
allowed to decrease if necessary in the readjustments. For
the final run of a series, the feed solution is diluted to about
10 L with water and sampled to determine the exact acidity and
the total curium-americium remaining in the feed. The actinide

content is used to size the final resin batch (7.7 mL of resin
per gram of actinide plus some excess to prevent breakthrough),
and the acidity is used to determine the final volume of feed.
The feed solution is then evaporated to approximately 1 L and
transferred to the head tank. The feed is diluted in the feed
head tank by a series of small water flushes that are made in
the feed adjustment tank to flush out all of the feed solution.
Six readjustments for a seven-run series have been handled in
this manner without problems.

Resin Loading. The resin has a capacity for curium and
americium of about 0.13 g/mL; therefore, 145 mL will sorb about
19 g of actinides. An excess of adjusted feed (typically 2.25 L)
is transferred to the feed head tank, then drawn through the
column at about 1.5 L/h (3.5 mL/cm^2·min). The loading progress
can be followed easily by darkening the hot cell and observing
the orange-red glow from ^{244}Cm if ^{244}Cm is present in sufficient
quantity, or less distinctly with the lights on by observing a
dark ring of radiation damage on the resin bed which is located
just above the loading front. The in-line alpha detector will
detect curium breakthrough but is usually not relied on for the
diversion of the raffinate stream when the curium glow can be
seen distinctly. As the loading front nears the bottom of the
resin bed, the effluent is diverted from the raffinate catch
tank to the feed adjustment tank to collect the curium-
americium-containing raffinate and the resin washes, thereby
eliminating loss of actinides from the feed system. The excess
feed is transferred through the column and is followed by two
small water flushes of the feed head tank and a final 200-mL
water flush that is added directly to the top of the column.
The resin is then allowed to dry for 10 minutes, while the pump
continues to pull air through the resin, before the column is
removed to the calcining furnace. Raffinate batches from the
individual loading runs are analyzed for ^{244}Cm and usually can
be discarded to waste.
   A final run of the series is made with as little as 3 g of
curium-americium remaining in the feed. For this partial size
run, the resin volume is sized with as much as 20 mL of excess
resin to prevent curium-americium loading breakthrough. This
change in the loading philosophy for the final run allows all
the actinides in the feed solution to be converted to oxides and
eliminates rework solutions from the feed system.

Calcination. The quartz column containing the loaded resin
is removed from the equipment rack and placed upright in the
cylindrical furnace with the top frit in place. The resin is
dried for 1.5 h at 150°C without a purge through the bed to
remove most of the remaining water, which tends to bridge
against an air flow and lift out the resin bed. After the initial drying period, the purge line is attached to the bottom of

the column, air flow is established at 0.3 L/min (superficial velocity of 7 mm/s) from outside the hot cell, and the resin is thoroughly dried in the air stream for 1 h at 150°C. After the drying period, the temperature is increased to 800°C and held during a 4-h period (including the heatup time) in the air stream to remove carbon. Then, sulfur is removed by changing (outside the hot cell) to a 0.3-L/min flow of 4% H_2-Ar purge gas while the temperature is maintained at 800°C for 4 h. The purge is then changed back to air at 0.3 L/min, and the heating is continued at 800°C for four additional hours to promote further oxidation.

At the end of the firing schedule, the cylindrical furnace is cooled, the quartz column is removed, and the actinide oxide microspheres are transferred by pouring into the platinum-lined crucible. The open-top crucible is placed in a muffle furnace and the oxide is sintered at about 1050°C for 16 h. Although there is no forced air flow during the sintering step, the approximately 10-mm-deep bed of oxide is exposed to air inside the furnace. After sintering has been completed, the oxide is cooled prior to the oxide characterization steps.

Curium-Americium Oxide Characterization. Large clusters of microspheres which have sintered together are broken apart and passed through a 210-μm-opening, stainless steel screen using vibration and a sliding weight on the screen. The free-flowing, dust-free product, consisting of microspheres and small clusters of microspheres, is characterized by noting the color, weighing the oxide on a triple-beam laboratory balance, and measuring the oxide volume ("tapped" for compaction) in a 10-mL glass graduate. A typical product, which is predominately curium oxide, is usually a dark gray color, indicating a mixture of CmO_2 (black) and Cm_2O_3 (white). Products which are predominately americium oxide have the same physical appearance and cannot be easily distinguished from curium oxide. The accepted stoichiometry of a typical curium oxide product is approximately $CmO_{1.71}$. The tap density, obtained from the weight and volume measurements, is compared with a minimum standard of 2.0 g/mL as the key indicator of the acceptability of the oxide for target fabrication. The actinide oxide is placed in a stainless steel storage container, and the ^{244}Cm content (2.832 W/g) is measured using an in-cell calorimeter. The ^{244}Cm content, the curium and americium mass analyses from the feed, and the weight of oxide are used to calculate the product purity, usually in the range of 85-93% actinide oxide. The major impurity is believed to be carbon since analyses for other possible impurities give minimal indications. A composite product sample made up of a small sample from each batch in a series of runs is taken for spark-source mass spectrometric analysis and a weight-loss-on-ignition test. The impurity analyses are used to verify that the calculated limits for heat production and neutron absorption in the HFIR are not exceeded.

Process Variables

The acid concentration of the feed solution is an important processing parameter. Acid concentrations in the range 0.01-0.70 $\underline{M}$ were investigated in the development tests. In each test, the curium sorbed on the resin was sufficient to produce acceptable oxide products. However, the acid concentration of the feed is maintained in the range 0.20 to 0.35 $\underline{M}$ in the production runs. In one of the earlier production runs at lower acidity, a precipitate formed in the feed solution. This was thought to be caused by an unknown contaminant, probably a phosphate species from an earlier solvent extraction step. In the production runs, the reduced actinide capacity of the resin is noticeable at the higher acidities. Convenient batch sizes and short loading times for the current scale of production are achieved with actinide concentrations of about 10 g/L, but actinide concentration is not considered an important variable. When a choice has to be made as sometimes occurs when the breakthrough loading raffinates are recycled to the feed tank, the acid concentration is held in the desired range, and the actinide concentration is allowed to decrease. Process operation and product quality are unaffected.

The sequence of drying and calcination steps is important to reliable operation and to the production of acceptable product. The initial air-drying period without purge removes most of the free water and avoids operational difficulties. The drying at 150°C with air purge thoroughly dries the resin and avoids shattering the beads with internally generated steam during subsequent heating. The calcination with an air purge at 800°C removes the carbon from the resin matrix. The sulfur content is then reduced to less than 1% during the 4% H_2-Ar purging at 800°C. This is a crucial step in producing oxide having adequate density and particle strength for use in fabricating HFIR targets. In test runs, the product tap densities were increased from 1.0 g/mL to values as high as 2.5 g/mL when 4% H_2-Ar was used. The calcination in air after the sulfur removal helps to oxidize Cm_2O_3 to CmO_2, a much more desirable product because of its higher density. (Americium is believed to have similar behavior.) Tap densities of greater than 2.0 g/mL are generally an indication of acceptable product for HFIR targets while lower tap densities, especially lower than 1.5 g/mL, indicate that the product will have unacceptably low crushing strength and will contain fines created from excessive particle breakage. The low density also makes it difficult or impossible to fit the desired amount of actinides into a target. Curium oxide microspheres with tap densities of 2.0-3.0 g/mL are estimated to be 40-60% of the theoretical crystal density by comparison with tap densities and mercury densities obtained for rare earth oxide microspheres prepared by comparable methods. Curium oxide microspheres with a density of only 30-40% of the theoretical crystal

density, are barely adequate to obtain the desired curium loading in the target.

Curium oxide particles formed from large resin beads, 200-
300 μm, show acceptable tap densities but have large internal
voids. Such particles are more fragile than solid particles and
are therefore less desirable. The voids decrease with decreasing resin size and do not exist in oxides made from smaller size
resins of 50-150 μm diameter. Resin beads averaging 60-80 μm in
diameter are now used to eliminate the internal voids and produce sintered oxide spheres (although not completely spherical)
of about 20 μm diameter. Although this is on the low side of
the desired range (20-210-μm diam), spheres of this size sinter
together at 1050°C and produce small clusters of spheres that
are sized to less than 210 μm in diameter by screening.

Typical tap densities from routine oxide production runs
are 2.9-3.1 g/mL within a 2.7- to 3.6-g/mL range. This is an
increase from the typical 2.3-2.4 g/mL within a 2.0- to 2.7-g/mL
range experienced in the development test runs. The increase is
not attributed to a single fundamental change in a process parameter but is believed to be a result of three factors: (i) using
a slightly smaller resin of more uniform size, (ii) loading the
resin to saturation (to column breakthrough), and (iii) the
ability to control and reproduce all parameters in the present
equipment.

Results

The production and oxide characterization data for a series
of curium-americium oxide production runs are presented in
Table I. The table includes the composite feed analyses, the
product data for each run, and a summary of the product data.
Totals and averages are presented to indicate performance even
though the products are not usually combined. Approximately
93.7% of the feed material was converted into product. The normal
losses of actinides from the product are the result of oxide
particles that stick to the product handling equipment. These
are not actual process losses because they are eventually
returned to rework.

The analysis of the composite sample from the same series
of curium-americium oxide production runs is presented in Table
II. The content of carbon, the major impurity, is inferred
rather than directly analyzed. The analyses of curium-americium
oxide products generally reflect the purity of the feed, except
for carbon and sulfur from the resin and a few potential corrosion products.

Summary

Resin-bead loading and calcination techniques have been
used to produce all curium and americium oxide feed material

Table I. Typical Curium-Americium Oxide Production Data

[Data from a series of runs made March 2-8, 1978]

Feed Composition	
Curium:	66.45 g	[37.4% ^{244}Cm, 0.4% ^{245}Cm, 53.2% ^{246}Cm, 1.4% ^{247}Cm, and 7.6% ^{248}Cm]
Americium:	17.39 g	[100% ^{243}Am]
Total Actinides:	83.84 g	

Individual Run Data

Run No.	Actinide Losses to Loading Raffinate (mg)	Oxide Product Measurements and Observations					
		Weight (g)	Vol. (mL)	Tap Density (g/mL)	^{244}Cm (g)	Actinide Oxide (%)	Sample Weight (g)[a]
54-CO-1	2.3	24.46	8.1	3.02	5.84	89.6	0.11
54-CO-2	5.7	23.54	7.6	3.10	5.58	88.9	0.11
54-CO-3	26.8	22.70	7.8	2.91	5.39	89.1	0.14
54-CO-4	18.2	22.93	8.6	2.67	5.30	86.7	0.12
54-CO-5[b]	12.8	5.09	1.9	2.68	1.18	87.0	0.09
TOTALS[c] (or weighted averages)	65.8	98.72[d]	34.0	2.90	23.29[d]	88.5	0.57[e]

a
Samples are removed before product measurements are taken.
b
Partial batch as final run of the series.
c
Individual batches are usually kept separate.
d
Product yield is 93.7%.
e
Composite sample analyzed for impurities - see Table II.

Table II. Typical Curium-Americium Oxide Impurity Levels

[Analysis of the composite sample from a series of runs
made March 2-8, 1978]

General Impurities[a]			Rare Earth Impurities[a]	
Element	ppm[b]	%[b]	Element	ppm[b]
C	c	10[d]	Gd	450
S	4500	0.45	La	450
Zr	200		Ce	300
Fe	150		Eu	150
Ca	75		Nd	45
P	75		Sm	30
Cr	30		Pr	20
Pb	30		Tb	15
Al	15		Dy	15
Si	15		Other(total)	150
Na	10			
Ti	10			
Zn	10			
Other(total)	25			

a
Analysis by spark source mass spectrometer.
b
Based on total product weight.
c
Analysis not applicable for carbon.
d
Carbon content is inferred from sample weight loss during
 ignition, typically 5-12%, and from the difference between
 the gross impurity level and the total of other impurities.

(about 2.2 kg) for HFIR targets since 1971. The process based on Dowex 50W resin has progressed from a series of test runs, through special production runs, to routine production in permanent equipment beginning in 1975. Key attributes of this process are its reliability, high yields, and ease of operation. The process is well suited for remote operation in hot cells, although some delicate handling of the oxide product is still necessary. Yields approaching 95% are routinely obtained, and only one unacceptable product has been generated during routine production operations. No problems have been encountered in fabricating targets from this oxide or in the subsequent irradiation of these targets.

The present scale of production of 150-250 g/yr supplies the present need and is comparable with the level of other chemical process operations at TRU. Since the annual production is accomplished in two 8- to 12-day periods, there has been no reason to consider further scale-up of the process. However, the rate of production could easily be doubled by simply adding a second set of calcination equipment.

Acknowledgement

    This research was sponsored by the Office of Basic Energy Sciences, U.S. Department of Energy, under contract W-7405-eng-26 with the Union Carbide Corporation.

Literature Cited

1. Ferguson, D. E. Nucl. Sci. Eng., 1963, 17, 435-437. Ten following articles describe details of the program.
2. King, L. J.; Matherne, J. L. Proc. 14th Conf. Remote Syst. Technol., 1966, 21-27.
3. Bottenfield, B. F.; Hahs, C. A.; Hannon, F. L.; McCarter, R.; Peishel, F. L. Proc. 14th Conf. Remote Syst. Technol., 1966, 172-184.
4. Chattin, F. R.; King, L. J.; Peishel, F. L. Proc. 24th Conf. Remote Syst. Technol., 1976, 118-129.
5. Collins, E. D.; Bigelow, J. E. Proc. 24th Conf. Remote Syst. Technol. 1976, 130-139.
6. King, L. J.; Bigelow, J. E.; Collins, E. D. "Industrial-Scale Production-Separation-Recovery of Transplutonium Elements," ACS Symposium 2nd Chem. Congress North American Continent, 1980.
7. Sease, J. D. "The Fabrication of Target Elements for the High-Flux Isotope Reactor," ORNL-TM-1712 (1967).
8. Burch, W. D.; Bigelow, J. E.; King, L. J. "Transuranium Processing Plant Semiannual Report of Production, Status, and Plans for Period Ending June 30, 1968," ORNL-4376 (1970).

RECEIVED December 19, 1980.

GENERAL SEPARATION AND RECOVERY METHODS

The Application of Pressurized Ion Exchange to Separations of Transplutonium Elements

DAVID O. CAMPBELL

Chemical Technology Division, Oak Ridge National Laboratory, Oak Ridge, TN 37830

One of the first triumphs of ion exchange chromatography was the separation and identification of fission product rare earths in the Manhattan Project in the early 1940s. Initial publication of this work was withheld until 1947 when nine papers from the Oak Ridge National Laboratory and the Ames Laboratory at Iowa State University appeared in the Journal of the American Chemical Society (1–9). The science of rare earth separations was indeed revolutionized. Separations that had taken years could now be done in about a day.

The transplutonium elements and the rare earths, or lanthanides, are so similar chemically that what is true for one group is generally true for the other. In practice, process development work is usually carried out with lanthanides, and frequently, all the solutions end up as analytical samples. Transplutonium elements, in contrast, are so valuable that the goal is the maximum yield of pure products. Accordingly, the methods and equipment developed with rare earth separations are applied directly to heavy actinide production separations. These may be quite small in scale, but this is "production" for some of these elements. Most of the development data that are suitable for theoretical interpretation, however, are acquired with rare earths. Fortunately, such data can be transferred to actinide separations with great confidence, as long as certain precautions are taken.

Two basic approaches are used to separate these elements, namely, elution development and displacement development chromatography. Both were defined in the original work; and in both, the separations are based primarily on differences in complexing of the trivalent ions by an organic reagent during elution through a column of strong (sulfonic) acid ion exchange resin. Displacement development is appropriate to larger-scale separations because larger column loading can be used and product concentrations are higher. Elution development is particularly suited to smaller-scale (even tracer) separations because the product bands can be completely separated from each other. The

0097-6156/81/0161-0189$05.00/0
© 1981 American Chemical Society

division between the two methods is not sharp, but it is probably in the vicinity of a few grams.

Since World War II there have been significant advances in three general areas. One is the accumulation of data for the interaction of these elements with a large number of diverse complexing agents, including distribution coefficients, separation factors, and complex stability constants. The result is that α-hydroxyisobutyrate is generally used for elution development separations, following the work of Choppin and Silva in 1956 (10), and a buffer of one of the polyaminopolycarboxylic acids (such as EDTA, DTPA, or NTA) is used for displacement development. Citrate, which was used in the original work for both approaches, is now only of historical interest.

The second advance was the use of a metal "barrier" ion such as Cu, Fe, or Ni which was demonstrated by Spedding, Powell, and Wheelwright (11, 12). The metal ion forms a stronger complex with the eluent than do the trivalent ions of interest, and thereby holds back or "retains" these ions. This alleviated several problems and contributed substantially to the usefulness of the process.

The third advance, and the primary subject of this paper, occurred in a completely different discipline, biochemistry. This was the development of dependable systems for high-pressure liquid chromatography during the 1960s. The motivation for this development was the need to separate a number of very similar materials in biological and medical research, such as nucleic acids. The usual ion exchange chromatographic methods were partially successful, yet inadequate. It was recognized that greater resolution was required to gain information about several key problems, and the overriding goal was improved resolution.

Reviews of the biochemical work generally start with Martin and Synge in 1941 (13) and then jump to the work of Cohn on nucleic acid separations by ion exchange chromatography in 1949 (14). It so happens that Waldo Cohn was coauthor of one of those original publications on rare earth separations in 1947. It was a logical approach to apply this new chromatographic method in his original field of interest, biochemistry, once the wartime priorities were suspended.

Heftmann (15) has attributed the application of ion exchange chromatography in the nucleic acid field directly to the rare earth separations work of Tompkins, Khym, and Cohn (1) and the high-performance pressurized ion exchange systems evolved later to provide greater resolution. The fascinating point is that the technology that grew out of this work came full circle after some twenty years, with the application of the pressurized systems to separations of the trivalent actinides, the homologues of the rare earths. This technology, which was reviewed in 1976 (16), contributes importantly to several papers at this symposium.

The road to better resolution was really obvious. It was to use ever smaller and more uniform ion exchange particles. The problem is that the small particles cause an extremely low flow rate under ordinary operating conditions. One way to circumvent the problem is to apply a high pressure at the column inlet. The factors that made a success of pressurized ion exchange were the development and commercial availability of dependable hardware such as pumps, valves, and fittings, which occurred generally in the 1960s, and then the integration of the components into practical systems and application of the systems to appropriate problems.

Actinide Production Considerations

The need for greatly increased production of the heavier actinides became apparent in the 1960s. There had been prior separations of multigram quantities of americium and curium, but only much smaller amounts of heavier actinides. Two programs were initiated. The High Flux Isotope Reactor (HFIR) and the Trans-uranium Processing Plant (TRU) were built at Oak Ridge National Laboratory (ORNL) (17), and they have continued to supply elements up to fermium, as discussed in other papers here. The Californium Production Program (18) was established at the Savannah River Laboratory (SRL) to produce ^{252}Cf in multigram quantities for market development.

Both these programs were designed originally to utilize solvent extraction predominantly, although at ORNL anion exchange was also scheduled for enrichment of the transcurium elements. Cation exchange was used for the final purification of the individual transcurium elements. At both sites various operation-al problems developed with solvent extraction, whereas ion exchange performance was unexpectedly good. In addition, pressur-ized ion exchange, which was developed at that time, permitted ion exchange to be applied to the highest radiation levels anticipated in these programs. Presently, ion exchange is used almost exclu-sively for the actual transplutonium element separations, and batch solvent extraction is utilized for removal of corrosion, activation, and fission products.

It was recognized that a very high radiation intensity would be encountered, well beyond that in reactor fuel reprocessing, for example. At the same time, the chemical separations are among the most difficult to accomplish. The separation factors for succes-sive pairs of transplutonium elements vary upward from about 1.3, and both the yield and the extent of separation from adjacent elements are desired to be about 99.9%. These requirements trans-late into a separation system capable of achieving about 500 equivalent theoretical plates. This is not really outstanding performance in terms of small-scale, modern, high-resolution chromatography; but it is exceptional performance when the radia-tion damage problems are taken into account.

The scale of work ranges from hundred-gram to kilogram quantities of combined fission products, rare earths, americium, and curium, down to 100-mg quantities of californium, <1 mg of einsteinium, and <1 pg of fermium. The radiation power density encountered in some separations is uniquely large for a chemical operation. There are several radiation damage effects, occurring in both the aqueous phase (where radiolysis causes gas generation) and the organic phase or resin (where resin properties are changed). When the programs were started, there was at least a little doubt about how the separations would be accomplished and how successful they would be.

The requirements for this work, then, may be generally summarized as reasonably good (but not extreme) resolution, small to moderate capacity (up to perhaps a few moles), suppression of the effects of radiation damage (particularly gassing), and a very high speed of separation to minimize the radiation exposure. These goals are uniquely met by pressurized ion exchange chromatography.

The short diffusion paths of the small resin particles provide good resolution with very high flow velocities, and the high pressure allows the high flow rate actually to be achieved. As a result, the elutions are fast, thereby diminishing the exposure time and, proportionately, the radiation damage. The high pressure, which is something of a necessary evil in conventional work, is a distinct benefit here because it alleviates the gassing problem. Gas solubility is proportional to pressure; so gases dissolve and, if bubbles do form (near the bottom of the column), they are immediately swept out.

Elution development

The first application of pressurized ion exchange to lanthanide and actinide separations was initiated in 1967 to examine this technique for the final separation of trivalent actinides in the TRU facility. In initial work with rare earths, it was demonstrated that 200-mg quantities of Nd and Pr could be adequately separated in times under an hour (19). This pair is as difficult to separate as any actinide pair, and the time and scale of separation easily met the requirements projected for actinide production at TRU. Subsequently, separations of multimilligram quantities of all fifteen rare earths were demonstrated in times as short as about 1.5 h (20, 21).

Mixtures of transplutonium elements were also studied after operating conditions and procedures were defined (Fig. 1) (22). This work used elution development with buffers of α-hydroxyisobutyric acid partially neutralized with NH_4OH, and Dowex 50W-X12 resin with a particle size range of 20—40 μm. The elution bands showed good separation of Es, Cf, Bk (the band labelled "β" is a combination of Bk and Eu), and Cm. The very small amount of Fm present in this early production material is in the extreme leading edge of the Es band.

Although the pressurized ion exchange equipment was patterned after systems being studied for biochemical separations, it was operated somewhat differently. For actinides, the speed of separation is critical because of radiation effects, and it was practical to trade off some resolution for a faster separation. In practice, much higher flow rates and somewhat larger resin particles were usually used. Flow velocities were typically in the range of 5 to 25 ml cm^{-2} min^{-1} {cm min^{-1}} for elution and up to 50 cm min^{-1} for loading. In contrast, conventional ion exchange separations of trivalent actinides commonly used 0.1 to 0.2 cm min^{-1}. Thus, the time required for the separation was reduced by factors of 10 to 100, with a comparable reduction in radiation damage.

System performance was excellent with 20–40 μm resin particles in columns up to 1.5 m long, but larger resin particles were often used to permit higher flow rates with available pressures, up to 10 MPa. A narrow range of resin particle sizes was beneficial, certainly within a factor of 2. Column loadings of 5 to 10% of the resin capacity gave good results, and problems with band overlap were significantly greater at higher loadings. The resin was uniformly packed into the column by loading it from a slurry with forced flow so that the flow velocity was substantially greater than the resin settling velocity. It was important that the column did not contain acid before the elution (sometimes appreciable amounts of acid were present in the feed, and this loaded on the column) because H^{+} ion would interact with the eluent and reduce the pH to an undesirably low value during the early part of the elution.

For processing batches containing higher radiation levels, it was beneficial to use a separate "loading" column, a short column usually of larger diameter. After loading the actinides at a very high flow rate to reduce radiation damage, the column was washed with dilute NH$_4$NO$_3$ solution to remove H^{+} ion and then valved to the top of the long elution column. In this way, most of the resin (which is in the elution column) was not exposed to impurities in the feed.

Band tailing is a problem in some of these separations because of the extreme variation in the amounts of different elements present. This is indicated by the elution curves for Es and Cm in Fig. 1. It is possible to follow these elements over concentration ranges of about 10^6, and it is characteristic that each band tails to the right (i.e., into the volume following the elution peak) at a concentration gradually decreasing from about 10^3 to 10^5 times smaller than the peak. The band tails contribute some degree of impurity to all succeeding bands; thus, in this system, lighter elements are contaminated to some extent by any heavier element present, since the normal elution order is heavy actinide to light. The practical significance of this is that bands following each element will be decontaminated with respect to that element by a factor about 10^4 or 10^5, whereas bands

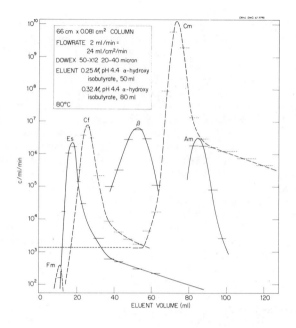

Industrial & Engineering Chemistry
—Product Design and Development

Figure 1.   Rapid separation of transplutonium elements (22)

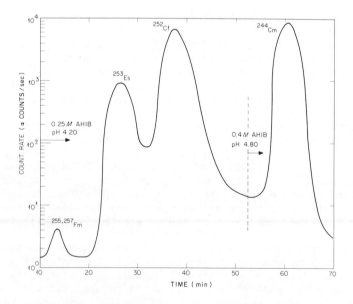

*Figure 2.   Elution curve given by flow-through α-detector during early pressurized
ion exchange separation of transplutonium elements (34)*

preceding can be more highly contaminated. Of course, further decontamination can be achieved by repeated cycles.

Production Separations

Because of its promise, the pressurized ion exchange approach was applied immediately to transcurium element production at TRU, and Fig. 2 indicates the sort of separation that was obtained. This shows the relative alpha count rate given by an in-line detector, and it demonstrates good separation of Fm, Es, Cf, and Cm. Berkelium is also well separated, appearing between Cf and Cm, but it is not shown because it is not an alpha-emitter.

Typical "plant" equipment is shown in Fig. 3. The equipment racks are assembled and tested before being installed in hot cells. This rack, which served for some five years, has one high-pressure pump, a short "loading" column, two long elution columns, and appropriate associated valves, feed vessels, product collection apparatus, and plumbing. Resin was periodically replaced by hydraulic transfer.

These columns are 1.2 m long and are made from stainless steel tubing up to 2.5 cm in diameter. The resin is graded into size ranges such as 25–50 or 70–100 μm diameter, the size selected depending on the application. Flow velocities are typically 15 cm min^{-1} for loading (they can be much higher if necessary) and 12 cm min^{-1} for elution. Full-scale separations of transcurium elements normally take less than 3 h; and second-cycle purification of individual elements and separations of two elements require less than 1 h. These methods are entirely adequate for all present and planned production requirements.

Elution development is used at TRU because the preliminary partial separation (LiCl anion exchange) yields a product containing all the transcurium elements along with only a small fraction of the Am, Cm, and rare earths. The total amount of trivalent elements to be processed in one run is generally not more than a gram. Burney and Harbour (23) demonstrated elution development separations with 35 g of Am plus Cm along with 1 mg of Cf. This work used a column 5.1 cm in diameter by 122 cm long and a flow rate of 8 cm min^{-1}. It is significant that elution development can be used for this scale of work, and it is doubtful that separations of such quantities would even be attempted without pressurized ion exchange.

The time average throughput for elution development with a pressurized ion exchange system is high, comparable to that with displacement development; and it should be practical to use this method for larger-scale separations. For separating larger quantities of short-lived alpha-emitters (i.e., ^{242}Cm or ^{252}Cf), elution development is superior, in principle, because the relatively high velocity of the elution bands and the low concentration of the elements in the bands (compared to displacement development) diminish the radiation damage effects.

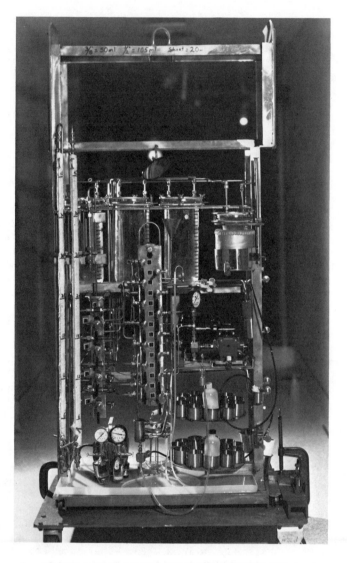

Figure 3. Typical equipment used for pressurized ion exchange separation of transplutonium elements (35)

Displacement Development

Larger-scale separations of americium and curium are based on
the pioneering work of Wheelwright and coworkers (24) and the very
large-scale use of displacement development for commercial rare
earth separations. Application of pressurized ion exchange to
displacement development for transplutonium element separations
has been pursued at SRL by Hale, Lowe, and coworkers (25, 26), and
the work was reviewed in 1972 (27).

With displacement development the metal complexes are eluted
at essentially a constant concentration (determined by the concentration of complexing agent and the pH), and the different elements are eluted sequentially one after the other. In
contrast to elution development, overlap regions between the bands
are an inherent characteristic of this method. If only a small
amount of an element is present, there may be no reasonably pure
product; rather, it will all occur in association with neighboring
elements. It is common practice to use several columns in series,
with the diameter decreasing sequentially, for example, columns
successively 10, 7.5, 5, and 2.5 cm in diameter. Since the bands
are longer in the smaller-diameter column, the overlap region
becomes relatively smaller, and a larger fraction of the product
is obtained in pure form.

One problem with this system is that Cf and heavier actinides
may not be retained by Zn^{2+}, which is the usual barrier ion; at
pH 6 these heavy actinides run ahead into the large volume of Zn-
DTPA solution that precedes the products. The result is an excellent separation from Cm, but a rather messy recovery operation
because of the large volume of solution from which the actinides
must be recovered. This problem can be overcome by using Zn at a
lower pH, such as 3 (27), or a different barrier ion, such as Ni^{2+},
but then the heavy actinides are found in the leading part of the
Cm band.

For separations involving large amounts of Am, Cm, or rare
earths, displacement development provides a satisfactory first-
cycle separation and yields Am and Cm products and a transcurium
element fraction suitable for final separation by elution development. However, alternative methods for the first cycle (removal
of the bulk of the lighter actinides and rare earths) are available; besides displacement development chromatography, these
include solvent extraction and the LiCl-anion exchange system.
The latter system is used at TRU, while the SRL development
program demonstrated the suitability of displacement chromatography. Both methods appear to be satisfactory. Until now, and
for the foreseeable future, the quantity of transcurium elements
has been too small to justify any process other than elution
development for the final separation.

With displacement development there is an intermixing of the
heavy rare earth fission products (above gadolinium) and the
transplutonium elements. Wheelwright demonstrated that Am and Cm

can be separated from the rare earths by using different complexing agents in successive cycles, namely, DTPA and NTA (24). The
elution order is shifted somewhat for the two complexing agents,
so that elements that interfere with Am or Cm when one complexing
agent is used do not interfere when the other is used. In this
way, quite pure products were obtained.

Kelley investigated a modification of the process to separate 1.7 kg of Am and 0.7 kg of Cm (a mixture containing 20 g of
^{242}Cm) from 167 moles of fission product lanthanides (28). This
material probably had the highest decay energy of any material to
be processed on that large a scale. Unusually high flow rates
were necessary to prevent boiling because of heat generation from
radioactive decay in the curium band. The last column in the
series (smallest diameter) has the greatest problem from heat
generation because the curium band is longest there; it was
necessary to elute this column with a flow velocity of 32 cm
min^{-1}. Full-scale separations were demonstrated using heavy rare
earth stand-ins for Am and Cm, but the system was not operated
with the actinides. However, this indicates the direction for
separations of larger amounts of highly radioactive actinides.

One possible application in which large amounts of rare
earths and actinides would be processed occurs in some schemes
for nuclear waste management. If it should prove to be advantageous to remove transplutonium elements from nuclear waste, for
example, the recovery of Am and Cm from the much larger amounts of
rare earths would be required. This problem has been investigated
by the author in tracer tests with rare earth mixtures typical of
fission products, using a heavy rare earth such as holmium as a
stand-in for Am and Cm (Fig. 5). It is clear that the bulk of the
holmium can be recovered in reasonable purity, and that the bulk
of the lighter rare earths is effectively separated from the very
small amount of heavy rare earths, Am, and Cm.

Extraction Chromatography

There are some applications for which different chemical
systems are advantageous, and the choice with ion exchange resins
is rather limited. Extraction chromatography offers a wide range
of possible systems, and the elution order can be changed and
even reversed. Baybarz and Knauer applied the pressurized ion
exchange method to this technique for separating ^{248}Cm from the
^{252}Cf parent (29). With the α-hydroxyisobutyrate ion exchange
system Cm is contaminated by the tail from Cf, which elutes
earlier; and extremely high-purity ^{248}Cm is desired. With
extraction chromatography using di(2-ethylhexyl)orthophosphoric
acid (HDEHP) on a porous glass support as the fixed phase and
nitric acid as the mobile phase, Cm elutes first, and decontamination factors (DF) greater than 10^{10} were obtained after two
purification cycles. The high DF is directly attributable to the
reversal of elution order.

Horwitz and coworkers investigated very rapid separations of
tracer actinides using small extraction chromatographic columns,
with the goal being separation of very short-lived products of
nuclear reactions (30). They demonstrated separations of Am and
Cm or Bk and Cf into pure fractions in one or a few minutes.
However, it is not clear that extraction chromatography is in-
herently superior to ion exchange in this respect, since small
column ion exchange experiments with rare earths also suggest
that separations can be obtained in about one min (31). Experimental studies have not yet really established the limit of
the tradeoff between resolution and speed of separation, or the
time (rather than the height) of an equivalent theoretical plate.

Schadel, Trautmann, and Herrmann compared the HDEHP extraction chromatography system to ion exchange elution development for
rare earth separations (32) and found the two methods to be about
equally effective under optimum conditions. From their data for
separating seven rare earths plus yttrium, they projected that all
the rare earths could be separated in about 20 min. However,
they did not include in their study those pairs most difficult to
separate.

This problem, the simultaneous separation of all the rare
earths, is of recurring interest (21). Recently, Qaim and coworkers carried out ion exchange studies which showed that the
height of an equivalent theoretical plate increased sharply with
larger column loadings in the case of the light rare earths, but
there was little effect with the heavy members (33). It appears
that such separations can generally be carried out in times of
the order of an hour using either extraction chromatography or
elution development ion exchange.

It must be recognized that extraction chromatography is
clearly superior to ion exchange in the choice of exchange
properties, because of the wide range of organic extractants
that can be used. However, the incentive for maximum speed of
separation in these problems requires the same pressurized chromatography approach, using very small support particles. It has
been adequately shown that performance in pressurized systems with
very high flow rates is satisfactory with several useful extractants; but, at the same time, the potential advantages have not
been fully realized.

Future Developments

Both production of and demand for transplutonium elements are
relatively stable, and pressurized ion exchange processes that
have been used for several years are entirely satisfactory for
these requirements. Three general areas can be visualized which
could require extension of these methods. These are: (a) tracer-
scale separations in the shortest possible time for very short-
lived isotopes in nuclear research; (b) separations similar to
those utilized now, but on a much larger scale, perhaps because
of radioactive waste processing or production of heat sources;

and (c) separations on a larger scale of such moderately short-
lived isotopes as ^{242}Cm or ^{252}Cf.

For all three areas the direction is already well established. Techniques that have been studied would be refined and
optimized for the particular problem. Pressurized ion exchange in
its present state of development permits successful processing of
materials 10 to 100 times more radioactive than does conventional
ion exchange chromatography. It is not clear what the ultimate
limits may be, but a significant extension of present technology
is clearly available if the need should arise.

Acknowledgment

Research sponsored by the U.S. Department of Energy under
contract W-7405-eng-26 with the Union Carbide Corporation.

Literature Cited

1. Tompkins, E. R.; Khym, J. S.; Cohn, W. E. J. Am. Chem. Soc.,
   69 2769.
2. Spedding, F. H.; Voigt, A. F.; Gladrow, E. M.; Sleight,
   N. R. J. Am. Chem. Soc., 1947, 69, 2777.
3. Marinsky, J. A.; Glendenin, L. E.; Coryell, C. D. J. Am.
   Chem. Soc., 1947, 69, 2781.
4. Spedding, F. H.; Voigt, A. F.; Gladrow, E. M.; Sleight,
   N. R.; Powell, J. E.; Wright, J. M.; Butler, T. A.; Figard, P.
   J. Am. Chem. Soc., 1947, 69, 2786.
5. Harris, D. H.; Tompkins, E. R. J. Am. Chem. Soc., 1947, 69,
   2792.
6. Ketelle, B. H.; Boyd, G. E. J. Am. Chem. Soc., 1947, 69,
   2800.
7. Spedding, F. H.; Fulmer, E. I.; Butler, T. A.; Gladrow, E. M.;
   Gobush, M.; Porter, P. E.; Powell, J. E.; Wright, J. M.
   J. Am. Chem. Soc., 1947, 69, 2812.
8. Tompkins, E. R.; Mayer, S. W. J. Am. Chem. Soc., 1947, 69,
   2859.
9. Mayer, S. W.; Tompkins, E. R. J. Am. Chem. Soc., 1947, 69,
   2866.
10. Choppin, G. R.; Silva, R. J. J. Inorg. Nucl. Chem., 1956,
    3, 153.
11. Spedding, F. M.; Powell, J. E.; Wheelwright, E. J. J. Am.
    Chem. Soc., 1954, 76, 612 and 2557.
12. Powell, J. E. Chapter 5 in "The Rare Earths," Spedding,
    F. H. and Danne, A. H., Eds.; Wiley, New York, 1961; p. 55.
13. Martin, A.; Synge, R. Biochem. J., 1941, 35, 1358.
14. Cohn, W. E. Science, 1949, 109, 377.
15. Heftmann, E. "Chromatography, Second Edition;" Reinhold,
    New York, 1967; p. 627.
16. Campbell, D. O. Separation and Purification Methods, 1976,
    5(1), 97–138.

17. Burch, W. D.; Arnold, E. D.; Chetham-Strode, A. Nucl. Sci. Eng., 1963, 17, 438.
18. Groh, H. J.; Huntoon, R. T.; Schlea, C. S.; Smith, J. A.; Springer, F. H. Nucl. Appl., 1956, 327.
19. Campbell, D. O.; Buxton, S. R. Ind. Eng. Chem., Process Design Develop., 1970, 9, 89.
20. Sisson, D. H.; Mode, V. A.; Campbell, D. O. J. Chromatogr., 1972, 66, 129.
21. Campbell, D. O. J. Inorg. Nucl. Chem., 1973, 35, 3911.
22. Campbell, D. O. Ind. Eng. Chem., Process Design Develop., 1970, 9, 95.
23. Burney, G. A.; Harbour, R. M. Radiochim. Acta, 1971, 16, 63.
24. Wheelwright, E. J.; Roberts, F. P.; Bray, L. A.; Ritter, G. L.; Bolt, A. L. Rep. BNWL-SA-1492, 1965.
25. Hale, W. H.; Lowe, J. T. Inorg. Nucl. Chem. Lett., 1969, 5, 363.
26. Lowe, J. T.; Hale, W. H., Jr.; Hallman, D. F. Ind. Eng. Chem., Process Design Develop., 1971, 10, 131.
27. Harbour, R. M.; Hale, W. H.; Burney, G. A.; Lowe, J. T. At. Energy Rev., 1972, 10, 379.
28. Kelley, J. A. Rep. DP-1308, 1972.
29. Baybarz, R. D.; Knauer, J. B. Radiochim. Acta, 1973, 19, 30.
30. Horwitz, E. P.; Bloomquist, C. A. A.; Delphin, W. H. 170th National Meeting of the American Chemical Society, Chicago, 1975.
31. Campbell, D. O.; Ketelle, B. H. Inorg. Nucl. Chem. Lett., 1969, 5, 533.
32. Schadel, M.; Trautmann, N.; Herrmann, G. Radiochim. Acta, 1977, 24, 27–31.
33. Qaim, S. M.; Ollig, H.; Blessing, G. Radiochim. Acta, 1979, 26, 59–62.
34. Baybarz, R. D.; Knauer, J. B.; Orr, P. B. ORNL-4672, 1973.
35. Baybarz, R. D.; Knauer, J. B.; Orr, P. B. ORNL-5672, 1973.

RECEIVED December 24, 1980.

Experience from Cold Tests of the CTH Actinide Separation Process

J. O. LILJENZIN, G. PERSSON, I. SVANTESSON, and S. WINGEFORS

Department of Nuclear Chemistry, Chalmers University of Technology,
S-412 96 Göteborg, Sweden

The CTH actinide separation process was developed as a possible means to reduce the expected long term dose to man from a geologic repository containing solidified radioactive waste from the reprocessing of spent nuclear fuel. The distribution data for the elements present in significant amounts in the high level liquid waste (HLLW) from a Purex plant, the general principles and the flowsheet have been described in detail elsewhere (1-7). A short recapitulation of the general features of the CTH process will be made to familiarize the reader with the process.

Figure 1 shows the extraction behavior of the actinide elements when a HLLW solution is contacted with a solution of 1 M HDEHP (di(2-ethylhexyl)phosphoric acid) in an aliphatic diluent (Nysolvin 75A). It is evident from this Figure that a good extraction of all actinide elements can not be obtained at any single acidity. This led us to the decision to first extract Pa, U, Np and Pu at 6 M nitric acid concentration, then reduce the acidity to slightly below 0.1 M and extract the remaining actinides, mainly Am and Cm, at this lower acidity. After the removal of Pa, U, Np and Pu it was found that the excess nitric acid could easily be removed by extraction with a 50% TBP (tri-n-butyl phosphate) solution in Nysolvin, thus avoiding some of the difficulties, such as slow reaction rates and crud formation, encountered when using formic acid for the destruction of the excess nitric acid. The CTH process can thus basically be divided into three fairly independent extraction cycles, see Figure 2. The order in which the two extractants are used is such that any minor carry-over of reagent from an earlier cycle to the next will not impair the operation of any cycle.

As a preparation for a small hot test, the process and necessary equipment have been extensively tested using simulated HLLW solution of the composition given in Table 1. These tests have proved very valuable. However, due to the high cost of many of the chemicals used to prepare the simulated waste they have also been fairly expensive. Some of the experience gained will be discussed below. The phrases "inactive", "low active" and "high

0097-6156/81/0161-0203$05.00/0

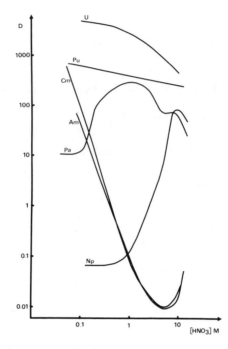

Figure 1. Distribution of actinides between synthetic HLLW solution and 1M HDEHP in Nysolvin 75A

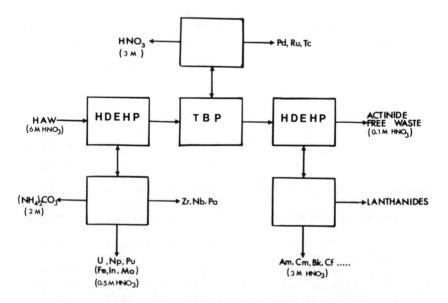

Figure 2. General outline of the CTH actinide separation process

Table 1. Composition, chemicals used and approximate costs of simulated HLLW solution employed for the cold tests. About 20 L of this solution is needed for one mixer-settler experiment.

Element	Conc. in HLLW g/L	Added chemical	Cost / 20 L batch $
Se	0.0068	SeO_2	0.06
Rb	0.13	$RbNO_3$	0.86
Sr	0.26	$Sr(NO_3)_2$	0.80
Y	0.17	$Y(NO_3)_3 \cdot 2H_2O$	15.3
Zr	1.28	$ZrO(NO_3)_2 \cdot 2H_2O$	10.1
Mo	1.18	$(NH_4)_6Mo_7O_{24} \cdot 4H_2O$	11.6
Ru	0.72	$RuCl_3 \cdot xH_2O$	771
Rh	0.15	$RhCl_3 \cdot xH_2O$	352
Pd	0.44	$Pd(NO_3)_2 \cdot 2H_2O$	408
Ag	0.022	$AgNO_3$	1.72
Cd	0.023	$Cd(NO_3)_2 \cdot 4H_2O$	0.08
In	0.0006	In_2O_3	0.13
Sn	0.0078	Sn	0.01
Sb	0.0032	Sb_2O_3	0.003
Te	0.040	Te	0.34
Cs	0.81	$CsNO_3$	23.0
Ba	0.61	$Ba(NO_3)_2$	0.59
La	0.43	$La(NO_3)_3 \cdot 6H_2O$	8.44
Ce	0.83	$Ce(NO_3)_3 \cdot 6H_2O$	12.3
Pr	0.41	Pr_6O_{11}	32.0
Nd	1.40	$Nd(NO_3)_3 \cdot 6H_2O$	45.5
Sm	0.27	Sm_2O_3	12.3
Eu	0.043	Eu_2O_3	37.5
Gd	0.039	Gd_2O_3	3.11
Tb	0.0008	Tb_2O_3	0.33
U	0.24	$UO_2(NO_3)_2 \cdot 6H_2O$	2.12
Fe	0.17	$Fe(NO_3)_3 \cdot 6H_2O$	0.86
Cr	0.034	$Cr(NO_3)_3 \cdot 9H_2O$	0.33
Ni	0.034	$Ni(NO_3)_2 \cdot 6H_2O$	0.09
PO_4	0.021	H_3PO_4	0.23
H	6.0	HNO_3	101.3

Total costs per 20 L , $ 1852

active" used refer to the expected conditions during the planned
hot operation.

Analytical procedures

The element by element analytical determination using
conventional techniques would be very laborious and time-
consuming for a solution as complex as the simulated HLLW.
Emphasis was therefore put on analytical procedures able to
determine many elements in parallel and/or requiring almost no
previous separation. The procedures preferred were X-ray
fluorescence using a ^{241}Am source and Si(Li)-detector, atomic
absorption spectrophotometry, gamma spectrometry using tracer
isotopes and Ge(Li)-detector and acid-base titrations with
recording of the pH-volume derivative. Table 2 summarises the use
of these methods for the different elements, and it also gives a
rough indication of interferences, sensitivity and accuracy
obtained.

Equipment

Mixer-settlers. To facilitate the testing of various
modifications of the flowsheet a modular mixer-settler unit was
developed, see Figure 3. As can be seen from this Figure, two
mirror mixer-settlers are used. These units can be put together to
obtain any desired number of stages in a battery. The sealing
between the mixer-settlers is made with nitrile rubber O-rings.
Each battery is begun and terminated with end-blocks, either
termination or connection-blocks. The termination-blocks are used
when there is no need to attach a second battery directly,
otherwise a connection-block is used. These end-blocks and mixer-
settler units are held together by five stainless steel rods
threaded in each end and tightened by washers and nuts.
The mixer is of the pump-mix type and able to pump the
aqueous phase up to a pressure difference of about 3 cm water and
the organic phase up to about 1 cm. The actual pumping effect is
of course dependent on the mixer speed, which is usually limited
by mixing and settling requirements. The impeller and shaft are
made from PVDF (polyvinylidene difluoride) and driven directly by
a small encapsulated 12 V DC motor fitted with a simple photo-
electric rotation-velocity transducer. The pump-mixer-motor unit
is fitted to the mixer chamber using a simple holder of plug-in
type, see Figure 3, in such a way that it can be retracted upwards
and replaced by remote handling should it fail during operation.
The motor current and velocity signals pass through an electric
connector at the top of the motor unit. Current to the motor is
supplied through separate cables from a motor control box, where
the speed and motor current can be read on a meter (0-3000 rpm /
0-3 A). The voltage applied to each motor can be adjusted
individually. Each motor control box can supply ten motors. The

Table 2. Analytical methods routinely used, major interferences, typical sensitivity and approximate accuracy for the elements present in the simulated HLLW.

Element	Method[a]	Major interference	Sensitivity (in HLLW) g/L	Accuracy (in HLLW) %
Se	XES	–	0.002	35
Rb	XES	Y,U	0.05	15
Sr	XES	–	0.005	2.5
Y	XES	Rb	0.01	10
Zr	XES	–	0.01	1
Mo	XES	Zr	0.05	2
Ru	XES	Mo	0.05	1.5
Rh	XES	–	0.005	6
Pd	XES	–	0.01	2.5
Ag	AAS	–	0.001	3
Cd	AAS	–	0.002	2
Sn	AAS	–	0.0005	5
Te	XES	Cs	0.003	7
Cs	XES	Te	0.005	1
Ba	XES	–	0.005	1
La	XES	–	0.01	4
Ce	XES	–	0.01	1.5
Pr	XES	–	0.005	2.5
Nd	XES	–	0.01	2
Sm	XES	–	0.01	3
Eu	XES	–	0.015	12
Gd	XES	Nd	0.03	50
U	^{235}U	–	0.01	2
Fe	AAS	–	0.005	5
Cr	AAS	–	0.001	5
Ni	AAS	–	0.01	5
H	pot. titr.	–	0.001	1

(a) XES = X-ray fluorescence spectroscopy with ^{241}Am source
AAS = Atomic absorption spectrophotometry

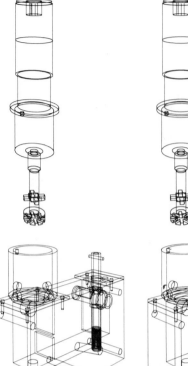
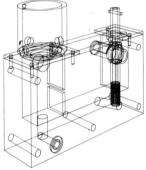
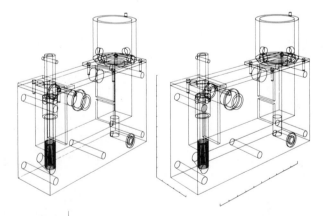

Figure 3.   Stereoscopic drawing of mixer and left and right mixer–settler building blocks; material used is plexiglass. In the lower right drawing three scales are shown, each having 1 cm between tick marks.

main causes of mixer failure have been commutator weardown and ballbearing corrosion.

The organic phase upper level is permanently fixed through an overflow weir in the end of the settler. This weir leads directly into the next mixer or into an end-block. A small adjustment is possible during assembly by rotation of the weirs. After assembly they are fixed in position by friction against the sealing O-rings.

The organic-aqueous interface level is controlled by a threaded vertical overflow tube in the aqueous outlet chamber, see Figure 3. This tube can be moved up or down by rotating it with a special manipulator that can be attached to the top of the tube. Because of the fixed organic phase upper level, changes in the aqueous outlet tube position leads to changes in the interface level. The aqueous phase passes down through the outlet tube and enters the adjoining mixer through its bottom supply hole or flows into an end-block.

The settler length in Figure 3, 7.6 cm, is the normal one used everywhere except in the 1G-battery, where a roughly three times longer settler is used to permit more time for phase disengagement.

Due to the small vertical dimensions of the mixer-settler only small density changes can be accommodated by the permissible vertical movement of the interface level. This implies that the interface level in each settler must be frequently determined and adjusted by means of the aqueous outlet tube position, especially during start-up of a battery or after changes in flowrates or solution composition. The interface position is measured by a high frequency (4 kHz) and low voltage (400 mV) conductance assembly, see Figure 4. To compensate for the small but unavoidable differences between different transducer-settler combinations, each level transducer is calibrated in the settler before start of operation by adding a fixed volume of aqueous solution, normally 75 mL, to each settler and then measuring the bridge unbalance voltage. This voltage is assumed to represent the signal corresponding to the normal position of the interface. As indicated in Figure 4, a small computer (CBM 2001) is used to switch a digital voltmeter (HP-3438A) to each level-transducer bridge in turn, read off the unbalance voltage and convert this to interface level position, which is continuously shown on a display and logged so that a history for the last 30 measurements of any settler interface level can be reviewed in graphical form when desired. This feature proved very valuable as a means to identify malfunctions in pump-mixers, feed-pumps or other equipment. Most of these show up as systematic changes in the interface levels. Figure 5 shows the pattern observed when a pump-mixer is malfunctioning and replaced.

To prevent organic phase flooding at low mixer speeds it turned out that a precise leveling of each battery is of major importance. Height differences between the ends of a 20 stage

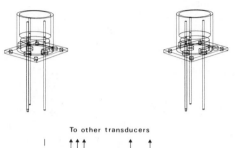

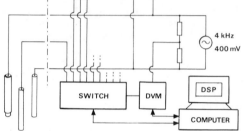

Figure 4.   Settler interface measuring device.   The upper part of the figure is a stereoscopic drawing of the level transducer.   Note that one electrode is covered with PTFE insulation, except for the lower tip.   Electrode materials have been graphite or titanium.   Lower part of the figure shows the level measuring circuit.

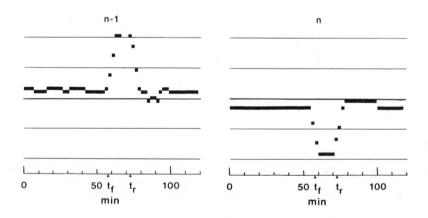

Figure 5.   Example of the displayed level history.   At time t_f a mixer is malfunctioning in stage n and replaced at time t_r.   Stage n-1 is the next stage in the direction of the organic phase flow.

battery as small as a few millimetres was found to be intolerable for this reason. It also turned out to be advantageous to have a small decrease in height between each battery in the direction of the organic phase flow. Usually a few millimetres was sufficient to facilitate organic phase flow between batteries.

Pumps. Liquid feedrates between 0.3 and 25 mL/minute are required for the process using the mixer-settlers described previously. Liquids were fed from supply tanks placed about 1.5 m below the batteries ("inactive solutions") or about 4 m below the batteries ("active feeds"). Different pump types were tested and an electronically controlled membrane pump with a PTFE membrane was selected as most suitable, being easily adapted to remote control, sufficiently corrosion resistant and giving flowrates in the desired range (Prominent Electronic A2001).

The valves supplied with these pumps showed an erratic behavior at low flow rates, mainly due to a too large diameter of the valve balls and poor design of the valve seats and valve chambers. New pumpheads and valves were therefore developed, manufactured and mounted on the pumps. The heads and fittings were made from PVDF with valve seats and 3 mm diameter balls in acid resistant steel. On these pumps the stroke length has to be set mechanically and this set-screw was modified so that it could be turned by a simple manipulator, if necessary. The electronic frequency control was modified for remote operation.

Flowmeters. Even after modification the pumps usually showed some long-time drift in the pumping velocity, mainly due to changes in valve operation caused by corrosion, dirt accumulation or trapped air bubbles. Thus all flowrates have to be measured at regular intervals and the pumps adjusted when needed. Various devices for flowrate measurements were tested and two were finally selected.

For all "inactive" or "low-active" liquids, the feed tanks were fitted with level tubes connected to the tanks by magnetic valves and in permanent connection with the pump feed lines. When the magnetic valve is closed, the pump takes its feed only from the liquid in the level tube. Knowing the inner diameter of the tube, the flowrate can be calculated from the time needed to lower the liquid in the graduated level tube by a given amount. Time is measured manually with stopwatches. After each measurement the magnetic valve is opened again, permitting free flow of liquid from the tank to the level tube and pump. The main difficulty with this simple device is due to the pulsation from the pump, giving a stepwise change in liquid level during the measuring period. This necessitated the measurement of both the level difference and time for an integral number of pump strokes. By using different diameter tubes for different streams it is possible to arrange so that all measurements can be done in less than one minute. The accuracy of these measurements is better than 5% in the flowrate.

For the "high-active" feed a different method for flowrate
measurement was selected. The feed flowrate is indirectly determined by measuring the flowrate out of the battery to the intermediate storage tank. The feed flowrate is then calculated from
this flowrate and the known aqueous wash flowrate, which being
"inactive" can be measured by the method described previously. To
measure the flowrate down to the tank, a small vessel (30 mL) is
fitted to the top of the tank. This vessel has an overflow tube
directely down into the tank for safety and is connected to the
tank by a magnetic valve in the bottom of the vessel. This vessel
is also fitted with three conductivity electrodes at different
levels. The signal from these electrodes is used to operate the
magnetic valve as follows. When the liquid level is below the
middle electrode the magnetic valve closes, and when the liquid
touches the upper electrode the valve opens. By knowing the volume
released into the tank for each valve cycle and the cycle time,
the flowrate can be calculated. The valve signal is monitored and
timed by the small computer used for settler interface measurements and the flowrate automatically computed and displayed after
each valve cycle. A count of the total number of cycles is also
kept and displayed as total accumulated volume in the tank. During
calibration runs this device showed a precision of better than
0.5 %. However, during long operation it was found that dirt and
crud could periodically prevent the magnetic valve from closing
completely, thus giving a too low flowrate. It was also found that
the weight of the liquid collected in the tank could lead to small
shifts in the tank position, e.g. a small tilting, which could
change the volume between the upper and middle electrodes somewhat
and thus introduce further errors in the flow measurement. To
minimize these effects a certain redesign of the device is desirable; (i) ensure better valve closure, (ii) separate the vessel
mechanically from the tank, (iii) increase the height to diameter
ratio of the vessel and (iv) filter the outgoing stream.

Pipes and fittings. All liquid streams are carried in PVDF
tubes, bent at 60 ^{o}C to the desired shape. This material has a
very good resistance against all solvents and chemicals used in
the process. It is also fairly radiation resistant (8).
For pumped streams, tubes of 2 mm inner diameter and 4 mm
outer diameter are used. For gravity flowing streams, tubes with
8 mm inner and 10 mm outer diameter are used. These tubes are
easier to use than conventional stainless steel tubes due to a
larger flexibility and smaller weight. They are also better then
PTFE tubes because PVDF has almost no tendency to flow under
pressure as compared to PTFE and PVDF has a far better radiation
resistance.
To permit a leakproof and secure connection between equipment
and the PVDF tubes a special type of fitting was developed and
tested, see Figure 6. This fitting uses a cone of PTFE as a primary sealing element. To ensure a fairly constant pressure on the

PTFE cone a nitrile rubber O-ring is used as a compressible element. The O-ring also provides a second barrier against leakage, should the PTFE cone fail. The experience from the use of these fittings is very favorable. No visible damage was observed after a 1.1 Mrad irradiation using a ^{60}Co source.

Tanks. Pyrex glass or high density (HD) polyethylene vessels are used for storage of "inactive" and nearly "inactive" solutions. The "high active" solutions are held in tanks of HD polyethylene fitted with the measuring vessel described above and also having three conductivity electrodes for indication of near empty and near full conditions. The margin from contact with the upper electrode to overflow is about 20% of the tank volume. The level indication for empty or filled tanks are shown on a small panel using LED-indicators. When any tank reaches a full indication an audible alarm is sounded, which can be reset manually. A similar alarm for the empty condition has not been used, but it is highly desirable to remind the operators that the end-of-feed condition is near thus allowing timely preparation for the switching back to inactive feed used for cleanup. It should also prevent the operators from erroneously assuming a failure of the "active" feed pump, piping or volume measuring device when the "high active" feed tank runs empty.

Distillation and evaporation. Ammonium carbonate solutions are used in several parts of the process; as stripping reagent in the first cycle and as organic phase clean-up reagent in the second and third cycles. Ammonium carbonate is recovered from the resulting solutions by distillation. The procedure used is slightly different for the two different uses of ammonium carbonate.

The most difficult part is in the recovery from the first cycle strip solution. This solution has a large tendency to foam and this property dominates the design of the whole distillation unit, see Figure 7. The recovery of the actinide strip solution is done in several steps.

Firstly the feed solution to the recovery unit is preheated to about 80 °C in a vessel with fixed volume (overflow back to feed tank). Then it is fed into the empty first evaporator. The design of this unit gives a reasonable distillation rate without overfoaming and it is fitted with an electrode which detects foam rising above a given level. Should this occur the heater is switched off for a fixed time, sufficient to let the foaming decrease, and then the heating is resumed. The evaporation is continued until a fixed condensate volume, sensed by a pair of level electrodes, is reached. During this operation, care has to be taken not to use too cold cooling water in the condenser as this will lead to the accumulation of masses of solid ammonium carbonate crystals in the condenser. Ultimately the condenser may clog or break.

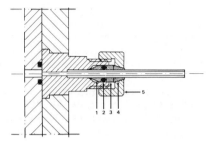

Figure 6. Drawing of fitting used for connection between PVDF tube and end-block: 1, PTFE sealing cone; 2, neoprene rubber O-ring; 3, PVDF washer; 4, PVDF retaining cone; 5, nut

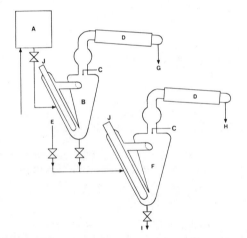

Figure 7. Drawing of ammonium carbonate recovery unit and evaporator; rated input capacity is 0.8 and 1.5 L/h with and without foaming respectively: A, metering tank with preheater; B, ammonium carbonate distillation, 0.5 kW; C, foam level sensor (conductivity); D, condenser; E, acid inlet, 6M HNO_3; F, evaporation unit, 1 kW; G, Ca 2M $(NH_4)_2CO_3$, to carbonate adjustment with CO_2; H, distillate, H_2O + NO_x; I, metal nitrate concentrate, ca. 0.5M HNO_3; J, electrical immersion heater

The concentrate contains a fine dark slurry consisting mainly
of actinide and iron carbonates. It is essentially free from
ammonium ions. This slurry is fed down into the partly empty
second evaporator and at the same time a fixed volume of 6 M
nitric acid is added. The amount of acid used gives an acidity of
about 0.1 M after dissolution of the slurry. When the transfer of
solutions is complete the connecting valves are closed, and the
liquid in the second evaporator is boiled until a fixed condensate
volume is collected. The concentrate is then partially removed to
a fixed level in the evaporator using an overflow weir. The
foaming in the second evaporator is much less than in the first
and the foaming decreases rapidly during the evaporation cycle.
Part of the mannitol present in the feed to the evaporators
precipitates in the second evaporator and at the end of the evaporation cycle a small amount of yellow crud (mannitol) is floating
on the surface. However, after an initial period, no further
increase in the amount of floating crud is observed. It is
presumably oxidized to soluble compounds.
   The ammonium carbonate solutions from the second and third
extraction cycles contain very little dissolved salts. These
solutions are retained in the second evaporator. This is only
emptied at the end of each experiment. Thereby the amount of
secondary waste is kept as low as possible.

   Sampling. The cover over each settler and aqueous outlet
chamber has a 6 mm hole. Samples of the phases are taken through
these holes using a remotely operated pipette and transferred to
small glass vials. The sampling volumes used are 0.1 or 1 mL.

Estimated dose to solvents

   The extraction of several radioactive elements, especially
the lanthanides and trivalent actinides, is commonly expected to
give doses to the solvents used in excess of those obtained in a
normal Purex reprocessing plant. To clarify this point, an approximate calculation of solvent doses from homogeneous irradiation
was made. The effects of inhomogeneous irradiation of the organic
phase from elements only present in the aqueous phase was
neglected as it was found to be about an order of magnitude
smaller than the effect from homogeneous irradiation.
   Larger amounts of highly radioactive elements are only present in the organic phase in the first and third cycles. Doses in
the TBP cycle are expected to be of the same order of magnitude or
slightly larger than in a Purex process. The main concern is
therefore about the doses to the HDEHP solution.
   During the aging of the HLLW solution from a Purex plant
insoluble precipitates are known to form, which could endanger the
operation of any actinide recovery process and increase actinide
losses. It is therefore believed that an actinide separation process must use the HLLW solution as soon as possible after it is

generated. The activity level of this solution is therefore
determined by the fuel type (BWR or PWR), its burnup and its
cooling time before reprocessing. It is therefore of interest to
estimate the effect of cooling time on the solvent dose. As a
standard case BWR fuel at a burnup of 27 000 MWd/te was chosen.
It was further assumed that mixer and settler holdup times were
the same as used in our small scale equipment. Mass transfer
between the phases in the mixer was further assumed to be
instantaneous to facilitate the calculations. In some cases this
may lead to underestimation of solvent dose, but in other cases
to overestimation. It is believed that these errors largely cancel
out over the whole process. It was further assumed that a small
common buffer tank was used for the purified recirculated solvents,
thus giving an averaging effect on the solvent doses in the first
and third cycles.

At cooling times less than 2 years the HDEHP mean dose is
dominated by extracted ^{90}Y, ^{91}Y and ^{144}Ce-^{144}Pr. At cooling times
from 2 years to 30 years the dose is mainly from ^{90}Y. After 60
years the dose from ingrown ^{241}Am dominates. The total dose to
the solvent as a function of cooling time is shown in Figure 8.
From this figure it is obvious that a delay of about 5 years
before reprocessing is beneficial for the actinide separation
process and that little is gained in terms of solvent dose by
further delays.

These estimated doses are believed to be correct within a
factor of two. They are not very large compared with the doses for
other applications of HDEHP reported in the literature (9) and
comparable with the doses for TBP reported for reprocessing of
high burnup fuel using mixer-settlers — 1.02 Wh/L at 43 000 MWd/te
and 50 days cooling (10).

Operation

Start-up. As mentioned earlier the settlers are first filled
with aqueous phase up to the normal interface level to permit a
new calibration of the interface level transducers. All aqueous
phase outlets are set in a higher position than normal to prevent
the unintentional escape of aqueous phase from any mixer-settler.
The mixer motors and organic feed pumps are then started, thus
filling each battery in turn with organic phase. At the beginning
these pumps are operated at higher than normal pumping speeds to
reduce the time needed to fill all mixer-settlers with organic
phase. When all batteries are nearly filled with organic phase the
pumping speed is metered and adjusted to the normal value that
shall be used during operation.

The aqueous feed pump to the last battery in the cycle (down
the organic stream) is started and adjusted to the prescribed flow
rate. A start at this end is necessary to guarantee that stripping
and clean-up batteries are operating properly before the organic
phase becomes loaded. When the interface level starts to rise in

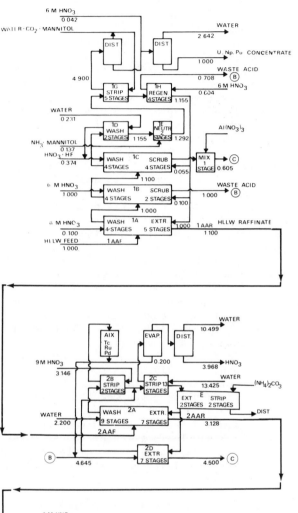

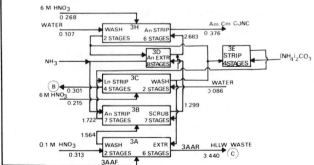

Figure 8. Estimated mean dose to the HDEHP solvent in the CTH process as a function of fuel cooling time

the settler nearest to the aqueous feed or wash its aqueous outlet
is adjusted downwards to keep the interface at the normal level.
The aqueous phase then flows into the next stage and the
adjustment is repeated for this stage and soforth throughout the
battery. When a battery is operating properly the procedure is
repeated for the next battery upwards the organic stream, until
all batteries are operating properly. Further adjustments may then
be needed on mixer speeds, typically 300-600 rpm, and the interface levels until a stable condition is reached.

    During the starting period, a simulated "high level" waste
from a separate tank is used as feed.

    Operation. "Active" operation is started by turning a three-
way valve in the feed-line to the "high active" feed pump. When
the new feed solution enters the system minor adjustments on motor
speeds and interface levels are usually necessary to compensate
for small changes in density and viscosity.

    Interface levels are monitored continously during operation.
The first sign of pump or mixer/motor malfunction or of crud accumulation is usually seen on the interface levels.

    Motor rotation velocities and currents are normally checked
and logged with a few hours interval. Pumping speeds are also
metered and adjusted with about two hours interval, keeping a
careful log of all data and changes made on the pumps.

    Routine sampling of the 1AAR, 2AAR or 3AAR streams, depending
on which cycle is running, are made with about two hours interval,
see Figure 9. For the first cycle, the 1AAR sample is analyzed for
iron, uranium and total α -activity as soon as possible. During
the first cycle, samples are also taken of the 1DAR stream and
analyzed for HNO_3 and HF using potentiometric titration in ethanol
with standardized NaOH. These data are needed for the adjustment
and make-up of the aqueous feed to the 1C-battery. During the
second cycle, the 2AAR sample is titrated to determine the concentration of free acid and during the last cycle, the 3AAR sample
is analyzed for total α -activity and titrated to determine the
concentration of free acid.

    Shut-down. When the end-of-feed condition occurs for 1AAF,
2AAF or 3AAF, the feed-line to the "high active" feed pump is
switched over to the "inactive" feed tank, which can contain
simulated HLLW solution, but more normally only HNO_3 of the
appropriate concentration. This solution is used until most of the
equipment has been roughly decontaminated. Then all feed-pumps and
mixer-settlers are shut off. The mixer-motor units and level
transducers are removed from the batteries and the mixer-settlers
emptied from solution using a hose and pump attached to a waste
tank.

    The run-down period required is about 15 h, 30 h and 20 h for
the first, second and third cycles respectively.

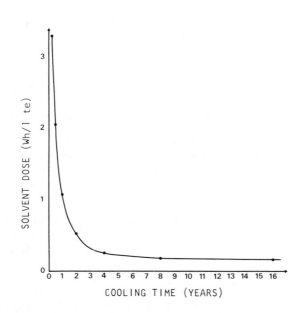

Figure 9. Flowsheet of process with nominal flowrates and with battery and stream names referred to in the text. In the experiments reported, the flow rates for the aqueous feeds have been: 1AAF 5 mL/min, 2AAF 1.5 mL/min and 3AAF 9.4 mL/min.

General observations

In general it has been found that the process, at least when divided into three separate cycles, is surprisingly easy to start, operate and stop. However, each of the cycles has certain critical points which needs special attention.

In the first cycle, the two mixer stages where the organic phase is mixed with ammonia-mannitol solution need special attention. The addition of the aqueous solution is controlled by a pH-electrode in the organic phase at the outlet from the second mixer stage connected to a titrator controlling the 1E-feed pump. A nominal "pH"-value of about 11 is used with the electrode calibrated in standard buffer solutions. The functioning of this electrode and the titrator-pump unit is crucial for proper phase disengagement in the 1G-battery. A "pH"-value which is too low gives poor phase separation. Also a wrong "pH"-value can lead to clogging of the two mixer stages, giving a reversal of the organic phase flow back into the 1D-battery. Should this occur, the operation of the 1D-battery is impaired due to evolution of carbon dioxide which results in heavy foaming. An early warning is obtained from the level indicators in the 1D-battery. Before any disastrous backflow of organic phase begins the interface levels in the 1D stages have fallen continously during a considerable time, maybe more than 30 minutes. In most cases a strong increase in mixer-motor speeds in the neutralization stages for a few minutes will remove the clogging. In some cases cleaning of the pH-electrode has been necessary to restore normal operation. In extreme cases the mixer-motors have had to be removed, and the mixers and overflow channels cleaned mechanically from accumulated precipitates.

In the second cycle, the aqueous and organic phase densities vary for each stage within the batteries. Any change in feed rates to the 2A- and 2C-batteries can lead to shifts in the interface levels which are large compared to the small settler height. In extreme cases some settlers have been completely emptied of one of the phases, leading to internal recirculation of the other phase to the previous stage. In such a case the level transducers do not give a proper reading of the interface level. It is always indicated as being too high. A careful study of the stored level history of such a stage will normally provide the information needed. If an aqueous level which is too high exists, then the level reading should have been rising continously. When an aqueous level which is too low exists the level reading has been falling continously and then suddenly jumped to the high indication.

In the third cycle the pH-control of the DTPA-lactic acid solution is critical. A pH-value which is too low anywhere will lead to deposition of DTPA-crystals in the settlers near that point. These crystals are quite large and dissolve only very slowly. The largest danger for DTPA-crystallization seems to be during the start-up period before proper circulation of the DTPA-solution through the 3B and 3D batteries has been established.

Acknowledgement

   We are grateful to Professor J.Rydberg for initiating this
work, to I. Hagström, B. Hjorth, T. Rodinsson, M. Svensson, B.
Alfredsson and L. Fridemo for skilful laboratory assistance and to
H. Persson, L. Båtsvik, C. Skoglund and L. Ohlsson for mechanical
and electrical construction work. This project has been supported
by The Swedish Council for Radioactive Waste and The Swedish
Natural Research Council.

Literature cited

1. Liljenzin, J. O.; Svantesson, I.; Hagström, I., "A Possible
   Solution to the Long-Time Storage Problem for High Level
   Waste"; CONF-761020, Tucson, 1976; p. 303.
2. Svantesson, I.; Hagström, I.; Persson, G.; Liljenzin, J. O.,
   "Distribution Ratios and Empirical Equations for the
   Extraction of Elements in Purex High Level Waste Solution,
   I: TBP"; J. Inorg. Nucl. Chem., 1979, '41, 383.
3. Svantesson, I.; Persson, G.; Hagström, I.; Liljenzin, J. O.,
   "Distribution Ratios and Empirical Equations for the
   Extraction of Elements in Purex High Level Waste Solution,
   II: HDEHP", J. Inorg. Nucl. Chem., 1980, 42, 1037.
4. Svantesson, I.; Hagström, I.; Persson, G.; Liljenzin, J. O.,
   "Separation of Am and Nd by Selective Stripping and Subsequent
   Extraction with HDEHP Using DTPA-Lactic Acid in a Closed
   Loop", Radiochem. Radioanal. Letters, 1979, 37, 215.
5. Liljenzin, J. O.; Persson, G.; Hagström, I.; Svantesson, I.,
   "Actinide separation from HLLW", Proc. Sci. Basis for Nuclear
   Waste Management, Boston 1979, in press.
6. Persson, G.; Liljenzin, J. O.; Wingefors, S.; Svantesson, I.,
   "Reducing the Long-Term Hazard of Radioactive Waste", Proc.
   2nd Technical Meeting on the Nuclear Transmutation of
   Actinides, EUR 6929, Luxenburg, 1980, p. 247.
7. Liljenzin, J. O.; Hagström, I.; Persson, G.; Svantesson, I.,
   "Separation of Actinides from Purex Waste", Proc. ISEC´80,
   Liege, 1980, 3, paper 80-180.
8. Skiens, W. E., "Sterilizing Radiation Effects on Selected
   Polymers", Radiation. Phys. Chem., 1980, 15, 44.
9. Schulz, W. W., "Radiolysis of Hanford B Plant HDEHP Extractant",
   Nuclear Technology, 1972, 13, 159.
10. Warner, B. F.; Naylor, A.; Duncan, A.; Wilson, P. D.,"A
   Review of the Suitability of Solvent Extraction for the
   Reprocessing of Fast Reactor Fuels", Proc. ISEC 1974, Soc.
   Chem. Ind., London, 1974, 2, 1481.

RECEIVED February 5, 1981.

Recovery of Americium–Curium From High-Activity Waste Concentrate by In-Canyon-Tank Precipitation as Oxalates

L. W. GRAY, G. A. BURNEY, T. W. WILSON, and J. M. McKIBBEN

E. I. du Pont de Nemours & Company, Savannah River Laboratory, Aiken, SC 29808

The Savannah River Laboratory (SRL) and Savannah River Plant (SRP) have been separating actinides for more than 25 years. Work continues to upgrade processes and to initiate new processes. This report summarizes work on a precipitation process developed to separate kg amounts of Am and Cm from hundreds of kilograms of $NaNO_3$ and $Al(NO_3)_3$. The new process includes formic acid denitration of the Am–Cm bearing streams for acid adjustment; oxalate precipitation of the Am–Cm; and Mn^{2+}-catalyzed HNO_3 oxidation of oxalate in both the decanted supernate and the precipitated actinides. The new process generates one-fourth as much radioactive waste as the solvent extraction process which it replaced and produces a cleaner feed solution for downstream processing to separate the Am and Cm before conversion to their respective oxides.

Origin of Am–Cm Solutions

Large-scale purification of [243]Am, [244]Cm, and [252]Cf by pressurized cation exchange has been planned at SRP for many years (1,2). Initial small-scale work involved isolation of a crude actinide-lanthanide fraction by batch extraction in the large (>10,000 L) SRP canyon tanks followed by solvent extraction and ion exchange in the SRL high level caves. Processing rates in the caves, however, were inadequate for large-scale purification of [243]Am–[244]Cm.

For large-scale purification, the Purex Plant solvent extraction bank was used first to separate Am–Cm from the Pu in the target element and then to separate Am–Cm from the Al in the target element. In each of the four campaigns that have been processed, the Pu in the target element has been purified by the normal Purex flowsheet. In each case, the Am–Cm fraction was initially rejected to the waste (1AW) stream.

0097-6156/81/0161-0223$05.00/0

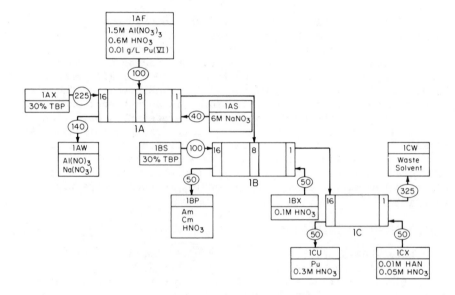

Figure 1.   Actinide separations flowsheet

Then for the first three campaigns, the 1AW was evaporated and acid adjusted to form feed for a second-pass through the solvent extraction cycle to extract the Am-Cm fraction. This flowsheet, shown in Figure 1, was used to recover a total of ~6.1 kg ^{243}Am and ~2.3 kg ^{244}Cm. The overall recovery of Am-Cm for these three campaigns was >99%. The purified product from these three campaigns was evaporated and combined in a single tank. Unfortunately, the contribution of both entrainment and solubility of NaNO$_3$ from the scrub (1AS) stream was sufficient to yield 8500 moles of NaNO$_3$ in the purified Am-Cm solution. The high-pressure ion-exchange process in the Multi-Purpose Processing Facility (MPPF), however, demands that the monovalent cation contamination of the Am-Cm feed be reduced to <620 moles. A further clean-up of this Am-Cm solution was therefore necessary to remove about 8000 moles of NaNO$_3$.

Since the first three campaigns did not produce a feed acceptable for downstream processing, the fourth campaign, producing about 8 kg of combined ^{243}Am-^{244}Cm, was stored after preparation of the 1AW stream as feed for a second-pass through the solvent extraction cycle. This then yielded the second solution, containing about 8 kg of combined ^{243}Am-^{244}Cm along with about 26,400 moles of Al(NO$_3$)$_3$. The high-pressure ion-exchange process, however, demands that the multivalent cation contamination of the Am-Cm feed be reduced to <31 moles. A further clean-up of the second batch of Am-Cm solution was therefore necessary to remove about 26,370 moles of Al(NaNO$_3$)$_3$.

An attempt was made to develop a precipitation process that would allow sufficient purification of both batches of Am-Cm to allow downstream processing in the MPPF to proceed.

Conceptual Process

It was necessary to use a process that would work in existing equipment, so a process was designed involving the following operations:

Acid Adjustment. To obtain low solubility losses of Am-Cm, the free acid (HNO$_3$ in solution plus HNO$_3$ available from hydrolyzable metal nitrates) concentration must be reduced to the 0.5-1.0M range.

Oxalate Precipitation and Digestion. Adjust the concentrations of Na$^+$, Al^{3+}, and H$_2$C$_2$O$_4$, and the temperature of the solution to grow sufficiently large crystals to allow the supernate to be decanted from over the crystals.

Decanting. Decant the solution using a steam-suction transfer system at the maximum possible rate and leave behind the last
possible supernate without transferring the precipitate.

Washing and Dissolving. Residual cation impurities are
removed by washing and decanting to dilute-out the heel supernate
before dissolving in the minimum volume of minimum concentration
HNO_3.

Oxalate Oxidation. Minimize the waste volume by oxidizing
the $H_2C_2O_4$ in the decanted supernate to $CO_2(g)$ and H_2O.

Am-Cm Finishing. Adjust the purified Am-Cm solution as
necessary for downstream processing.

Laboratory Demonstrations and Results

Each step of the process was determined on a laboratory
scale using Dy^{3+} as a surrogate for the lanthanides and actinides.
Initial precipitations of prepared surrogate solutions were performed in centrifuge cones to allow fast separation of the precipitate from the pregnant liquor. Simulations of in-canyon-tank
precipitation-digestion-settling were performed in a glass tank 14
cm in diameter and 81 cm high. The settling rates of various
digestion cycles and the volume of slurry produced were then
measured. Decanting was demonstrated by vacuum transfer of supernate using various size tubes and adjusting the vacuum to simulate
various steam-jet transfer rates. Open beakers were used to simulate product slurry dissolution and oxalate destruction procedures; off-gas rates from oxalate oxidation were measured by water
displacement from a water-sealed reaction train.

Formic Acid Denitrations. Simulated solutions were subjected
to laboratory formic acid denitrations (Figures 2 and 3). The
most usable free acid concentration for the simulated solutions
was obtained when a formic acid to free acid ratio of about 1.6 to
1.9 was used. This ratio yielded a final free acidity of about
0.6 to 0.8M. As a result of Al^{3+} hydrolysis, it was possible to
drive the Al-Am-Cm solution to about pH 10. However, acid concentrations less than 0.2M had to be avoided to prevent hydrolysis
and precipitation of the actinides.

As a result of the high concentration of nitrate (from sodium
and aluminum nitrate) the reaction rate was controlled by the formic acid addition rate until the free acid concentration was
reduced to about 0.5M. For semi-batch denitrations it appears
that a nitric acid concentration of 1 to 2M at the end of each
individual denitration is an excellent stopping point. Using 1 to
2M HNO_3 as a projected stopping point assured that there will be
no residual formic acid at the end of the reflux and evaporation

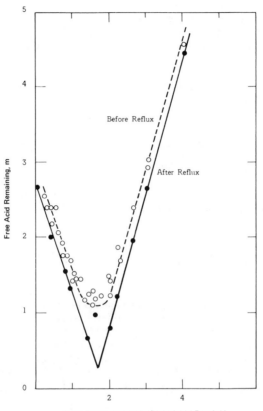

Figure 2. Formic acid denitration of simulated Am–Cm–NaNO₃ feed solution

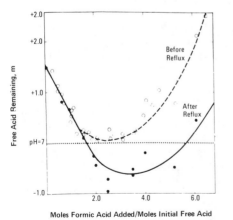

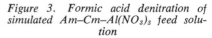

Figure 3. Formic acid denitration of simulated Am–Cm–Al(NO₃)₃ feed solution

step.   Additional high nitric acid–aluminum nitrate solution can
be added to the evaporated–denitrated solution without auto-
initiation of a formic acid–nitric acid reaction.

After all the solution has been transferred to the denitration evaporator and evaporated, it is possible to drive the denitration reaction to a residual free acid concentration of less
than 0.5M as required for the precipitation step.

Precipitation of Simulated Solutions.   For the Am–Cm–NaNO$_3$
solutions, acceptable losses (<1%) of transplutonium elements
could be achieved using 0.3M H$_2$C$_2$O$_4$ in the final slurry with a
free nitric acid concentration of <0.7M. These conditions were
achieved by adjusting the free acid concentration to <1.0M and
adding one volume of 0.9M H$_2$C$_2$O$_4$ to two volumes of adjusted feed.

As the result of oxalate ion complexing of Al^{3+}, precipitation of Am–Cm–Al(NO$_3$)$_3$ solutions was not straightforward. Using
Dy as a stand-in for Am–Cm, simulated solutions were prepared
where the ratio of Al(NO$_3$)$_3$ to Dy(NO$_3$)$_3$, KF, NaNO$_3$, and Hg(NO$_3$)$_2$
was held constant as would result in actual process solutions.
However, the total ratio of these species to free nitric acid was
varied in the stock solutions. Precipitation conditions were
simulated by additions of either a half-equal or an equal volume
of either an 0.9M H$_2$C$_2$O$_4$ or a saturated ($\sim$2M) potassium oxalate
solution. After percipitation and centrifugation, the residual Dy
in solution was determined by flameless atomic absorption. The
percent Dy remaining in solution was calculated (Figures 4 and 5).

These tests indicate that to obtain high yields from precipitation, the aluminum concentration of the slurry must be <0.2M
(Figure 5). This can best be obtained by denitrating the Am–Cm–Al
solution concentrate with formic acid (Figure 3) to about 0.5M
HNO$_3$. Then, dilution to an aluminum concentration of <0.5M would
yield a feed suitable for oxalate precipitation.

Digestion and Settling Rates of the Precipitate.   To design
the proper short leg-suction-jet, it was necessary to know the
volume of slurry to be expected in the tank, rate of settling, and
the minimum distance above the slurry that the jet must be to
prevent movement of the slurry.

The volume of precipitate and settling rate were determined
by precipiration of 4 L of simulated NaNO$_3$ solution preadjusted to
1M HNO$_3$ by the addition of 2 L of 0.9M H$_2$C$_2$O$_4$. The settling rate
(Figure 6) for the major portion of the precipitate was about 3.2
cm per min. Fines, however, settled at about 2.0 cm per min. The
volume of precipitate-slurry was determined to be $\sim$3.6 L per mole
of dysprosium.

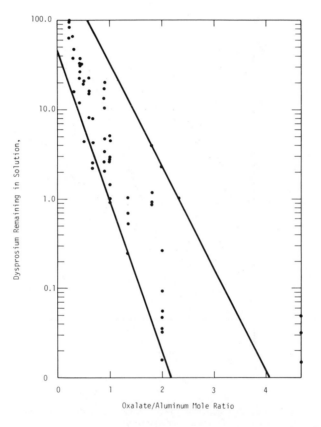

Figure 4. Precipitation of dysprosium from simulated waste as a function of oxalate ion concentration

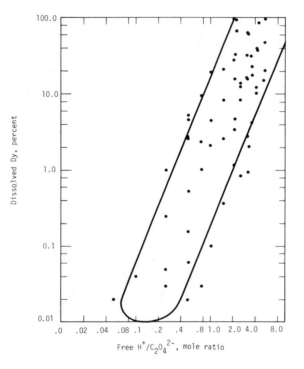

Figure 5. Precipitation of dysprosium from simulated waste as a function of free acid concentration

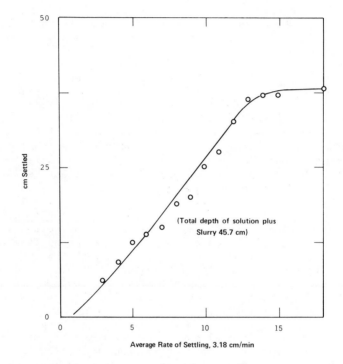

Figure 6. Settling rate of typical simulated oxalate precipitate

Projecting these data to a 12.5 m^3 tank indicates an approximate 2-hour settling time for the major portion of the precipitate and perhaps 4 to 6 hours for the remainder of the fines to settle. Digestion of the precipitate at $\sim$40°C for about 2 to 4 hours, however, eliminates fines.

The volume of precipitate and settling rate were also determined by precipitation of 2 L of simulated solution, adjusted to $\sim$0.5M Al^{3+} and $\sim$0.25M HNO$_3$ by the addition of 4 L of 0.9M H$_2$C$_2$O$_4$. When the precipitation was carried out at room temperature, less than 10% of the precipitate had settled after a 60-hour settling period. When the oxalic acid was added to a 60°C solution and then held at 40-45°C for an additional 2 hours, the settling rate and final volume of precipitate were very similar to the Am-Cm material containing NaNO$_3$.

Washing the Precipitated Oxalates. As a result of the 2500-L heel of slurry left in the precipitation tank, the product slurry contained large concentrations of contaminating cations. To yield an acceptable product for downstream processing in MPPF, these ions had to be diluted from the product. Five equal volume washes of the slurry were calculated to reduce the non-lanthanide impurity concentrations of polyvalent cations to acceptable levels; only four washes are needed to reduce the monovalent cations to acceptable levels as shown in Tables 1 and 2.

Jetting of Waste Supernate from Over the Precipitated Slurry. To simulate jetting of waste supernate, solutions were transferred by vacuum from the simulated canyon tank to a second tank through a 0.64 cm ID tube. At linear velocities through the tube of about 9.6 m per min, solution could be transferred at about 3.81 cm above the slurry without moving precipitate. Movement of the transfer-tube closer than 2.5 cm resulted in the movement of precipitate into the tube.

The Purex canyon normally uses two standard transfer jets; one at 284 L per min, the other at 95 L per min. The face velocity of the 284 L per min jet is about 14.0 m per min; the 95 L per min jet is about 4.7 m per min. The 95 L per min jet was used at 15.2 cm above the slurry as an additional safety factor to prevent excessive transfer to precipitate.

Dissolution of Washed Precipitate. Downstream processing of the Am-Cm product requires that the precipitate be dissolved, the oxalate ion removed from the solution, and the acid adjusted to <1M. Attempts were made to dissolve the precipitate at various nitric acid concentrations. Dissolutions in 1 to 5M HNO$_3$ were successful only if Mn^{2+} was added to catalyze the oxidation of oxalate ion. At a concentration of 8M HNO$_3$, the precipitate

TABLE I. Composition of $Na^+-Am^{3+}-Cm^{3+}-NO_3^-$ Solutions

	Storage Condition Before Purification	MPPF Feed Requirements	Actual Final Solution
Volume, L	17,000	<620	
HNO , M	2.72	<0.5	
Moles, total	46,420	<310	
NaNO , M	0.50	<1.0*	
Moles, total	8,500	<620*	45
Ln's + Ac's, total moles	∿550	∿550	
M	∿0.032	∿0.887	
Other polyvalent cations, M	∿0.015	<0.05**	
Moles, total	∿255	<31**	13

* This is the feed requirement for the sum of all polyvalent cations (Al^{3+} + Fe^3 + Hg^{2+} + all others).

** This is the feed requirement for the sum of all monovalent cations excluding H^+ (Na^+ + K^+ + any others).

TABLE II. Composition of Al^{3+}-Am^{3+}-Cm^{3+}-NO_3^- Solutions

	Storage Condition Before Purification	MPPF Feed Requirements	Actual Final Solution
Volume, L	13,200	620 L	
HNO_3, M	1.0	<0.5	
Moles, total	13,200	<310	
$Al(NO_3)_3$, M	2.0	<0.5*	
Moles, total	26,400	<31*	50
$NaNO_3$, M	0.07	<1.0**	
Moles, total	925	<620**	75
KF, M	0.12		
Moles, total	1584		
$Hg(NO_3)_2$, M	0.023		
Moles, total	304		
$Fe(NO_3)_3$, M	0.023		
Moles, total	304		
H_2SO_4, M	0.046		
Moles, total	607		
Ln's + Ac's, M	0.006		
Moles, total	80		

---

* This is the feed requirements for the sum of all polyvalent cations (Al^{3+} + Fe^{3+} + Hg^{2+} + all others).

** This is the feed requirement for the sum of all monovalent cations excluding H^+ (Na^+ + K^+ + any others).

could be dissolved at temperatures of 60 to 80°C. Because down-stream processing requires that the extraneous polyvalent cation to lanthanide-actinide ratio must be less than 0.14, addition of Mn^{2+} must be kept to a minimum. Because the precipitate can be dissolved in 8M HNO_3 without the addition of Mn^{2+}, 8M HNO_3 should be used to dissolve the precipitate.

Oxalate Destruction in Product Stream. As the Am-Cm will be separated by an ion exchange process, it is necessary to remove all of the oxalate ion. If separated solution is to be stored for an extended period of time before separation, self-radiolysis of the solution will of course destroy a portion of the oxalate. Mousty, Toussaint, and Godfrin (3) have shown that extended boil-ing of a 10M HNO_3 solution oxidized sufficient oxalate to render a suitable solution for separation. However, extended boiling of >10M HNO_3 solutions in stainless steel leads to excess corrosion of the equipment and, hence, the introduction of polyvalent cations (Fe, Cr, Ni) to the process solution. Koltunov (4) has shown that manganous ion (Mn^{2+}) will catalyze the oxidation of oxalate in nitric acid solutions. Low concentrations of manganous ion were therefore used in an attempt to catalyze the oxidation of oxalate in the precipitated slurries.

Figure 7 shows the destruction of oxalate in the precipitated slurry in 8M HNO_3. A manganous ion concentration of 0.045M Mn^{2+} in 8M HNO_3 will oxidize the oxalate in the slurry in about one-half the time required at a concentration of 0.011M Mn^{2+}.

Because Mn^{2+} is a polyvalent cation, its concentration in the final solution feed to the cation columns should be kept below 0.05M to avoid excess competition for resin sites. For the over-all process, it is better to use the longer oxidation times than to use higher Mn^{2+} concentration. Catalyzed oxidation should not be performed until the volume of solution is reduced to the mini-mum possible volume.

Oxalate Destruction in Waste Stream. Although the reactions are more rapid at 8M HNO_3, manganous ion catalyzes the oxidation of oxalate at lower acidities. Approximately 1M HNO_3 seems to be required. As the major solids producing reagent in the waste stream is oxalic acid, it is much cheaper to oxidize the oxalic acid to CO_2 gas than to store it as radioactive waste. Waste streams are therefore acidified with nitric acid, $Mn(NO_3)_2$ added and the solutions evaporated. During the evaporation, the oxalate ion is oxidized to CO_2 gas.

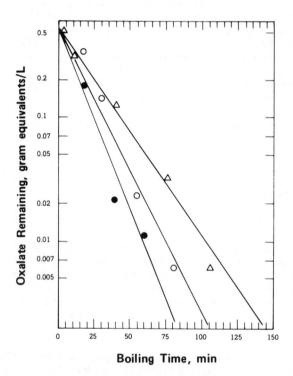

Figure 7.   Oxidation of oxalate ion with Mn²⁺-catalyzed 8M HNO₃: (●) 0.045M Mn²⁺, (○) 0.022M Mn²⁺, (△) 0.011M Mn²⁺

Flowsheets. Generalized flowsheets for the separation procedures are given in Figures 8 and 9. The flowsheets provide for
acid adjustment by formic acid denitration followed by dilution
and precipitation in batches with precipitated slurry accumulated
in the tank. After washing of the precipitated slurry to remove
the contaminating cations, the product is dissolved by adjusting
the acid concentration to <8M and heating the solution. Product
is then stored until the MPPF is available to process the material.

Supernates are transferred to an evaporator containing boiling 5M HNO_3 and 0.02M Mn^{2+}. Solution is transferred at a rate to
maintain approximately a constant evaporator volume. Sufficient
acid is added at the end of each batch to oxidize oxalate to be
transferred in the next batch. All material is accumulated in the
evaporator.

When MPPF processing begins, the product solution must be
evaporated from the ⅋8000 L storage volume to about 600 L. If
oxalate has survived both the initial heating steps and the high
radiation field during storage, it must be oxidized during the
evaporator step. Sufficient Mn^{2+} should be added to give approximately 5 moles of Mn^{2+} in the MPPF evaporator. After evaporation
the acid must be adjusted to MPPF requirements by formic acid
denitration.

Summary of Plant Processing

Approximately 8 kg of Am–Cm were recovered from the stored
Am–Cm–$NaNO_3$ waste. The process yielded an overall recovery of
94.75% of the Am–Cm while rejecting ⅋99.5% of the Na, ⅋95% of the
SO_4^{2-} and ⅋85% of the Fe. Of the losses, 0.25% represented
soluble losses and 5% was entrained losses.

Approximately 6 kg of Am–Cm were recovered from the stored
Am–Cm–$Al(NO_3)_3$ waste. The process yielded an overall recovery of
⅋75%. Of the losses, ⅋2.5% represented soluble losses and the
remainder was entrained losses. Of this 20% lost to the waste
supernate stream, about 14% (⅋1.1 kg) is stored for future
recovery; the remaining 7% was actually lost to the waste tanks.
The major contaminants of the purified Am–Cm product are Fe, Al,
and Na.

Characterization of Actual Solutions. The analyses of the
concentrate solutions before purification are given in Tables I
and II.

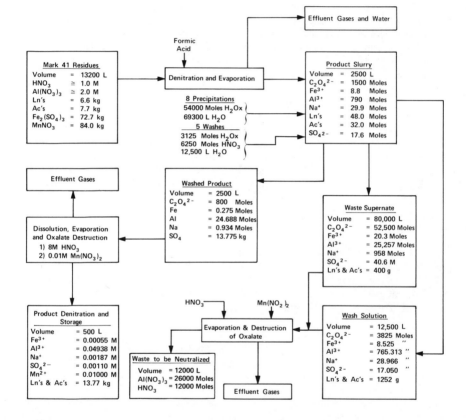

Figure 8. Flowsheet for the purification of Mark 41 residues

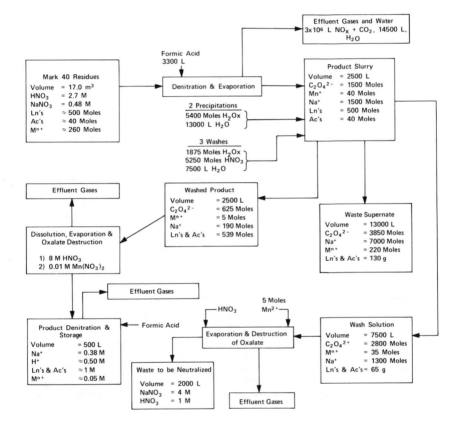

Figure 9. Flowsheet for the purification of Mark 40 residues

Formic Acid Denitrations. As a result of the denitrator tank
volume limitations, approximately 65% of the Na-Am-Cm-NO$_3$ solution
was transferred to the denitration tank, evaporated by 50%, and
partially denitrated with formic acid. Then the remainder of the
Na-Am-Cm-NO$_3$ solution was transferred into the tank, evaporated to
2500 L, and the complete solution denitrated to 0.9M HNO$_3$.

As a result of Al(NO$_3$)$_3$·9H$_2$O solubility and the denitrator
tank volume, it was necessary to split the Al-Am-Cm-NO$_3$ solution
into four batches for denitration. After denitration, the individual batches were recombined to yield a uniform Al-Am-Cm-NO$_3$
solution for precipitation. The final HNO$_3$ concentration of the
Al-Am-Cm-NO$_3$ solution was 0.2M.

Foaming was a problem with the Al(NO$_3$)$_3$ solutions due to the
high salt content; this extended both the denitration runs and the
evaporation period before denitration.

Precipitation, Digestion, Washing, Decanting of Am-Cm-NaNO$_3$
Solution. To minimize precipitation of sodium oxalate salts and
to minimize the solubility of actinide oxalates, the denitrated
solution was diluted with water and heated to ∿60 C before adding
oxalic acid. Sufficient 0.9M oxalic acid was added to adjust the
final slurry to 0.6M oxalate, 0.18M nitric acid and <0.5M sodium
nitrate. The slurry was cooled to 40-45°C and digested for four
hours to promote large crystal growth. After cooling to 35°C, the
precipitate was settled for eight hours before decanting the
supernate.

This procedure was repeated with the remainder of the
Am-Cm-NaNO$_3$ solution and four wash cycles. Losses to the waste
supernate were 5.25%; only 0.25% of the total was dissolved
losses, the remaining 5% loss was the result of entrained solids.

Precipitation, Digestion, Washing and Decanting of Am-Cm-
Al(NO$_3$)$_3$ Solution. As a result of oxalate complexing of Al^{3+}, it
was necessary to divide the Am-Cm-Al(NO$_3$)$_3$ solution into ten
batches. After transfer of about 10% of the solution into the
precipitation vessel, dilution water was added to reduce the Al^{3+}
concentration and the solution was heated to about 60°C. A small
amount of La^{3+} carrier was also added to the first batch before
sufficient oxalic acid was added to make the solution 0.6M. The
slurry was then cooled to 40-45°C and digested 4 hours to promote
crystal growth before decanting the supernate. This procedure
was repeated with the remaining nine batches until all the precipitate accumulated in the tank. The accumulated precipitate was
washed four times to dilute out contaminating cations.

Decanting losses were much higher than anticipated. In an attempt to minimize losses, the chemical composition of the precipitation batches was varied to increase the oxalate/aluminum ratio. These variables had little effect on product losses because the solubility losses were very low (<15% of the actual loss). Apparently, Am-Cm oxalate solids are being suspended prior to and during decanting (perhaps due to the decay heat of Cm) and a portion of the suspended precipitate was decanted. A new decant jet, which had a plate welded across the bottom and holes drilled around the pipe above the plate to provide horizontal rather than vertical suction, was fabricated and installed after the fifth batch. Losses, however, remain unchanged.

A total of about 20% of the product ($\sim$1600 g Am-Cm) was entrained to the waste evaporator with decanted supernate. Most of the oxidized supernate concentrate containing 1150 g Am-Cm was retained and is available for recovery by another method; the remainder was discarded.

Destruction of Oxalate in Waste Stream. Before supernate or wash solution was added, 5M HNO_3 containing 0.02M Mn^{2+} was brought to boiling in the evaporator to ensure prompt and rapid oxidation of the oxalate. Transfer rates of either supernate and wash solution or additional nitric acid, needed to ensure complete destruction of the oxalate, were limited to keep off-gas rates less than 28.3 m^3 per min. The final oxalate concentration of the waste was <0.02M which indicates >99.8% of the oxalate was oxidized to CO_2 and H_2O.

Dissolution and Characterization of Final Products. The washed product slurry was adjusted to 8M HNO_3 and heated to about 40°C to dissolve the oxalate precipitate. A comparison of the major impurities before and after the oxalate precipitation purification step for the sodium nitrate solution is given in Table I, and for the aluminum nitrate solution in Table II. These two solutions have been combined and will be stored to await processing by high pressure ion exchange in the MPPF.

Acknowledgement

The information contained in this article was developed during the course of work under Contract No. DE-AC09-SR00001 with the U.S. Department of Energy.

Literature Cited

1. Thompson, M. C.; Burney, G. A.; McKibben, J. M., p. 515 in
   Actinide Separations, ACS Symposium Series 117, 1979, American
   Chemical Society (1980).

2. Orth, D. A.; McKibben, J. M.; Prout, W. E.; Scotten, W. C.,
   p. 354 in Proc. Intern. Solvent Extraction Conference, 1971,
   Society of Chemical Industry; London (1971).

3. Mousty, F.; Toussaint, J.; Godfrin, J., "Separation of
   Actinides from High Activity Waste. The Oxal Process,"
   Radiochem. Radioanal. Lett., 1977, 31, 918.

4. Koltunov, U. S., "The Kinetics and Mechanism of Oxalic Acid
   Oxidation by Nitric Acid in the Presence of Divalent Man-
   ganese Ions," Kinetikai Kataliz, 1968, 9, 1034.

RECEIVED February 17, 1981.

Separation of Curium-242 From Irradiated Americium-241 Targets

WU KEMING, JIANG FASHUN, WANG RUIZHEN, CHEN MINGBO, WEI LIANSHEN, ZHUANG REHJIE, FAN YUANFA, CHEN HENGLIANG, ZHU RONGBAO, and QIAO SHENZHONG

Institute of Atomic Energy, Acdemia Sinica, Beijing, People's Republic of China

^{242}Cm is one of the most important isotopes of curium. It is an alpha-active nuclide with a half-life of 164 days and a specific activity of 3,300 curies per gram ($\underline{1}$).

To obtain weighable quantities of ^{242}Cm, ^{241}Am is irradiated with neutrons in a reactor. At the same time, plutonium, mainly ^{238}Pu and ^{242}Pu, and fission products are also formed ($\underline{2}$). Therefore, for the separation of ^{242}Cm from irradiated Am targets, the separation process has to include the separation of transplutonium elements (TPE) from other actinides, from lanthanides and other fission products. Since the 1960's several countries have developed processes for the separation of ^{242}Cm from irradiated ^{241}Am targets ($\underline{2}$-$\underline{5}$). In recent years we have also developed a process which can be used to extract hundreds of curies of ^{242}Cm from several grams of irradiated ^{241}AmO$_2$. In this process liquid-liquid extraction, extraction chromatography and ion exchange are used.

Recovery of Transplutonium Elements

Separation of Pu From Am and Cm by HDEHP

Separation of Pu From Am and Cm by HDEHP - Plutonium(IV) can be extracted by HDEHP (di-(2-ethylhexyl) phosphoric acid) from nitric acid of medium concentration; trivalent TPE and some other impurities are not extracted.

The dependence of extraction distribution coefficients on the concentration of HDEHP and HNO$_3$ and also the influence of impurities such as sodium, aluminum, iron and fluorine have been studied in our laboratory. These experiments show that if the concentration of HDEHP in kerosene is kept between 0.1 and 0.5$\underline{M}$ and HNO$_3$ kept between 2 and 3$\underline{M}$ the distribution coefficient of Pu(IV) will exceed 10^3 and the separation factor of Pu and Am will exceed 10^3. Under these conditions impurities described above do not affect the extraction of Pu(IV).

0097-6156/81/0161-0243$05.00/0

For stripping Pu from HDEHP, Fardy recommended an organic reducing agent (6). Our experiments prove that Pu(IV) can be quantitatively stripped by oxalic acid from HDEHP, while extracted zirconium remains in the organic phase. Usually, a decontamination factor of 200-300 for Zr/Pu can be obtained.

From our tracer experiments, the distribution coefficient of Pu(IV) is directly proportional to $[HDEHP]^2$ and is inversely proportional to $[H^+]^2$. The capacity experiments show that there are four molecules of HDEHP and one plutonium ion in a molecule of extracted compound. From this, we conclude that HDEHP is a dimer in the concentration range studied. The proposed extraction mechanism is as follows:

$$Pu(IV) + 2NO_3^- + 2(HDEHP)_2 = Pu(NO_3)_2(DEHP)_2(HDEHP)_2 + 2H^+$$

Separation of TPE and REE by Extraction Chromatography - Extraction chromatography was used for the separation of trivalent TPE and REE. HDEHP was used as the stationary phase on red-kieselguhr as the inert support. Trivalent TPE and REE were loaded on the column filled with this stationary phase from dilute nitric acid. Then, the column was eluted with a solution of 0.05M DTPA-1M lactic acid at pH=3.0. Trivalent TPE were eluted from the column. The influence of aluminum ion was studied, and the pH and flow rate of elution solution and column temperature were also studied as a function of separation.

We found that absorption of Am^{3+} is greatly affected by aluminum nitrate. Trivalent Am can be satisfactorily absorbed from 0.1M HNO_3, but it can not be absorbed from 0.1M HNO_3 and 0.5M $Al(NO_3)_3$. In 0.1M HNO_3 and 0.01M $Al(NO_3)_3$ solution, the volume distribution coefficient of Am^{3+} is five times less than in 0.1M HNO_3. Therefore, if there is a great quantity of Al^{3+} in the feed solution, it should be removed.

The capacity experiments showed that even though two kinds of inert supports are made from the same material, they differ greatly in absorption capacity because of differences in preparation methods. The specific surface areas of white and red-kieselguhr are $1m^2/g$, and $4m^2/g$, respectively, so that the best capacity of the former is equal to 0.12 meq. per gram of staionary phase only, but the capacity of the latter is more than 0.72 meq. per gram of stationary phase. When the column is saturated with trivalent Eu the molecular ratio of HDEHP and Eu^{3+} is equal to three, indicating an absorption reaction of:

$$Eu^{3+} + 3HDEHP = Eu(DEHP)_3 + 3H^+$$

This equation agrees with the results reported in reference(7).
But it differs from the reaction in liquid-liquid extraction
which is:

$$Eu^{3+} + 3(HDEHP)_2 = Eu (DEHP)_3(HDEHP)_3 + 3H^+$$

From our experiments, the optimum processing parameters are:

Stationary phase	= HDEHP
Inert Support	= Red-Kieselguhr
Absorption solution	= 0.05-0.1$\underline{M}$ HNO_3-0.005M $Al(NO_3)_3$
Elution solution	= 0.05$\underline{M}$ DTPA-1$\underline{M}$ lactic acid, pH=3.0
Temperature	= 30°C
Flow rate	= about 2mL/min/cm^2

The separation factor of TPE achieved is 10^3 under these
conditions.

Separation of Am and Cm by High-Pressure Ion-Exchange - Ion
exchange chromatography is one of the most efficient methods
for the separation of trivalent actinides. Since the alpha
activity of ^{242}Cm is very high, it is advantageous to adopt
high-pressure ion-exchange techniques because they reduce
operation time and decrease the damage of resin by radiation.

Cation exchange elution chromatography has been used as
our separation method for Am and Cm. The separation conditions
have been studied in detail. They involved the dependence of
elution solution pH, temperature of the column and resin cross-
linking on resolution. When cross-linking of the resin equals
4 per cent and column temperature equals 50° to 75°C, resolution
is optimized.

With the increase of the volume of feed solution, the ratio
of H^+/NH_4^+ in the resin phase increases and the distribution
coefficient of Am and Cm also increases. But separation of Am
and Cm becomes poorer(8). In order to improve this, after
absorption a washing stage using 0.2$\underline{M}$ NH_4NO_3-0.015$\underline{M}$ HNO_3 is
added. In this way, H^+/NH_4^+ in the resin phase is adjusted to
its proper value.

The following processing parameters were selected:

Cross linking of cation exchange resin = 8%
Elution solution = 0.040$\underline{M}$ alpha-AHIB, pH = 3.84
Temperature = 50°C
Linear flow rate = 15 cm/min.

Reprocessing of Irradiated $^{241}AmO_2$ Target - Based on the
above results, a process of separation of ^{242}Cm from irradiated
$^{241}AmO_2$ target has been developed. Process diagrams are given
in Figures 1-3.

Preparation of Target, Irradiation in Reactor and
Dissolution - AmO_2 powder and an excess of Al powder were well
mixed and irradiated at a neutron flux of 6.8 x 10^{13} neutron/
cm^2/sec until an integral flux of 2.6 x 10^{20} neutrons/cm^2 was
achieved. The resulting mixture was dissolved as shown in
Figure 1.

Extraction of Pu - For the extraction of Pu the acidity
of the dissolver solution should be adjusted to 1.5$\underline{M}$ HNO_3,
with the addition of $NaNO_2$ to ensure complete conversion of Pu
to Pu(IV). Detailed processing parameters are shown in Fig. 1
The Pu was further purified by anion exchange.

Aluminum Elimination - 50 percent TBP in kerosene was used
to extract trivalent TPE and REE quantitatively from dilute acid
and higher salt solution, whereas Al^{3+} and the fission products
of alkali and alkaline earth metals remain in the aqueous phase.
Detailed parameters of this stage are presented in Fig. 2

Separation of TPE and REE - Fig. 2 presents the flow-
diagram of extraction chromatography. The column was filled
by stationary phase which was made up of HDEHP and red-kieselguhr
with a mesh size of 120-140 mesh. The column was 50 cm high and
the diameter was 2.5 cm. All operations were carried out at
28°C and the flow rates were controlled at below 2 mL/min/cm^2.

After absorption, the red-orange fluorescence of ^{242}Cm
could be seen clearly in the dark.

All operations from dissolution to extraction chromatography were carried out in hot-cells. Composition of nuclides
in the dissolver solution is shown in Table I. The distribution
of alpha-activity in hot-cell processing stages is given in
Table II. Practically, the distribution of alpha-activity is
just that of alpha-activity of ^{242}Cm because almost all alpha-
activity comes from ^{242}Cm. As shown in Table II, the yield
of ^{242}Cm is high enough to warrant hot-cell operations.
Table III shows the results of the purification of Am, Cm, and
Pu from fission products.

DTPA Elimination - Before separation of Am and Cm with
high-pressure ion-exchange, Am and Cm solutions containing
DTPA from extraction chromatography separation need to be

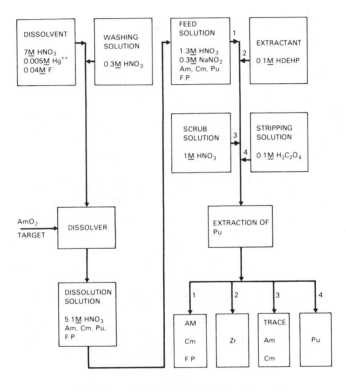

Figure 1. Dissolution and extraction of Pu(IV)

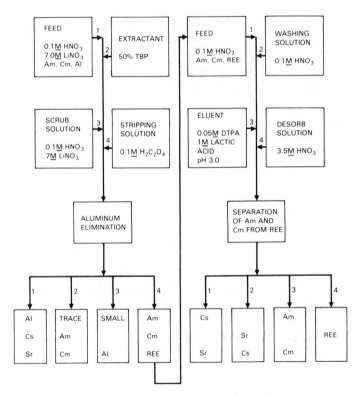

Figure 2.   Aluminum elimination and separation of Am and Cm from REE by extraction chromatography

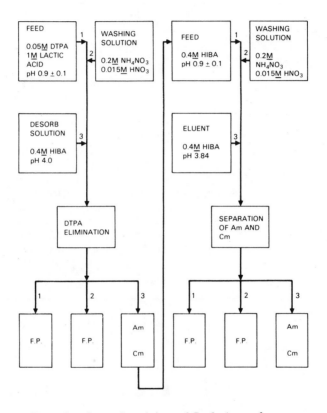

Figure 3. Separation of Am and Cm by ion exchange

TABLE I

COMPOSITION OF NUCLIDES IN DISSOLVER SOLUTION

Nuclide	^{242}Cm	^{241}Am	^{238}Pu	^{141}Ce	^{144}Ce
Activity (Ci)	148	2.20	0.30	0.61	0.33
Nuclide	^{103}Ru	^{106}Ru	^{95}Zr	^{95}Nb	^{140}La
Activity (Ci)	0.83	0.44	0.66	0.44	0.11

TABLE II

DISTRIBUTION OF ALPHA-ACTIVITY IN PROCESS SOLUTIONS

	Total Alpha-Activity (Ci)	Alpha-Activity/ Alpha-Activity Dissolution Solution (%)
Dissolver Solution	148	100
HDEHP Extraction Raffinate	148	100
TBP Extraction Raffinate and Scrub Waste	2.4	1.6
HDEHP Stripping Solution	0.008	Small
Effluent of Extraction Chromatography	Small	Small
Scrub Waste of Extraction Chromatography	146	98.6
Eluate of Extraction Chromatography	140	94.6

TABLE III

PURIFICATION OF Am, Cm, AND Pu FROM FISSION PRODUCTS

		DECONTAMINATION FACTOR				
		Total	^{141}Ce And ^{144}Ce	^{103}Ru And ^{106}Ru	^{95}Zr	^{95}Nb	^{140}La
Pu	HDEHP Extraction	4.9×10^2	2.7×10^3	1.2×10^3	2.7×10^2	2.9×10^2	--
Am And Cm	HDEHP Extraction	1.3	1	1	7.2	1.6	--
	TBP Extraction	7.9	26	3.3×10^2	7.1	89	46
	Extraction Chromatography	≥ 10	$\geq 1.3 \times 10$	≥ 7	≥ 7	≥ 17	

TABLE IV

DETECTORS FOR IN-LINE MONITORING

	Size	Application
Co-axial Ge(Li) Detector	35 Cm^3	Detecting γ
Stilbene Crystal Detector	40 x 40	Detecting Fast Neutron From ^{242}Cm Spontaneous Fission
Na(Tl) Thin Crystal Detector	50 x 1	Detecting 59.6 KeV γ Of ^{241}Am
Si(Au) Surface Barrier Detector	8	Detecting α

converted to the alpha-AHIB solution system. Conventional
cation exchange can serve for this purpose and Fig. 3 shows
the processing parameters. The yield of ^{242}Cm was more than
99.5 per cent. The Am and Cm solution was concentrated seven
fold in this stage.

Separation of Am and Cm - Fig. 3 shows the processing
conditions of high-pressure ion exchange for the separation
of Am and Cm. The column system consisted of an absorption
column and two separation columns, loaded with resin of 45-60
24-36, and 12-33 micrometers diameter, respectively. This
column system was kept at a constant temperature of 50°C.
The solution was fed into the column by a pump, which raised
the pressure to 40-70 kg/cm^2, to give a linear flow rate of
9 cm/min.

The column system was connected to an in-line monitoring
arrangement. Table IV presents the detectors used for in-line
monitoring. Eluate curves were drawn on the recorders. An
eluate curve of separated Am and Cm based on Ge(Li) detector
is shown in Fig. 4. With this in-line monitoring arrangement,
pure products of Am and Cm were collected.

Conventional ion-exchange and high-pressure ion-exchange
were carried out in a thick wall glove-box.

The final product of ^{242}Cm was examined by Si(Li), Ge(Li),
and α-γ coincidence spectrometers, the ^{242}Cm α-activity being
99.99 per cent, and ^{241}Am α-activity being less than 3 x 10^{-3}
per cent. The overall yield of the process is greater than
95 per cent.

Conclusion - The process described above is feasible for
the separation of several hundred curies of ^{242}Cm from several
grams of irradiated ^{241}Am target. Since HNO$_3$ is used as the
aqueous phase, stainless steel may be used as the processing
material. Techniques such as batch extraction, extraction
chromatography and high-pressure ion-exchange with in-line
monitoring are well adapted for this purpose.

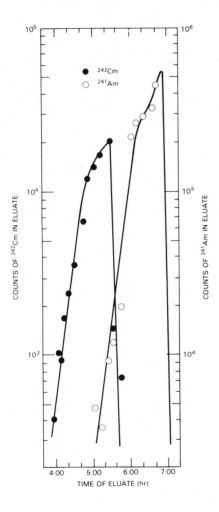

Figure 4. Elution behavior for separation of Am and Cm ion exchange

Acknowledgement

    Chen Yaozhong, Zhang Zefu, Pan Baolong, and Tang Quanyong
participated in this work.

Literature Cited

1. Keller, C., (1971) "The Chemistry of the Transuranium
    Elements," p. 620, Verlag Chemie Gmbh, Weinheim/
    Bergstr, Germany.

2. Vaughen, V.C.A.; McDuffee, W. T.; Lamf, E.; Robison, R.A.;
    Nucl. Appl. 1969, 6, 549.

3. Hohlein, V.G.; Born, H.J.; Weinlander, W.; Radiochim. Acta
    1974, 21, 167.

4. Horwitz, E.P., 1969, ANL-7569.

5. Specht, S.; Schutz, B.O.; Born, H.J.; J. Radioanal. Chem
    1974, 21, 167.

6. Fardy, J.J.; Chilton, J.M.; J. Inorg. Nucl. Chem.
    1969, 31, 3247.

7. Sochacka, R.J.; Siekierski, S.; J. Chromatography 1964, 16
    376.

8. Campbell, D.O.; Separation and Purification Methods 1976,
    5, 97.

RECEIVED April 17, 1981.

Precipitation of Actinide Complex Anions with Cobalt(III) or Chromium(III) Complex Cations

T. ISHIMORI, K. UENO, and M. HOSHI

Japan Atomic Energy Research Institute, 2-2-2 Uchisaiwai-cho, Chiyoda-ku, Tokyo 100, Japan

Actinide ions form aqueous-soluble complex anions with many kinds of ligands (1,2). Some of these actinide complex anions are precipitated as crystalline compounds by adding Co(III) or Cr(III) complex salts(3,4,5,6). Figure 1 shows, for example, the crystalline habit of the thoriumoxalato precipitate, $[Co(en)_3]_8$-$[Th(C_2O_4)_6]_3 \cdot nH_2O$.

The precipitates obtained so far by our Japanese research group are summarized in Table I. (The abbreviations used are as follows: en, ethylenediamine; tn, trimethylenediamine.) From these results, the following systems are considered to be applicable for the recovery and separation of actinides.

Carbonate System

Addition of ammonium carbonate to a solution containing an actinide(III), (IV),(V) or (VI) ion gives the following results. Only actinide(VI) ions form soluble carbonato complex ions. Actinide(III) and (IV) ions precipitate as their hydroxides or basic carbonates, and actinide(V) ion precipitates as a double carbonate. Therefore, in dilute ammonium carbonate medium, U(VI) ion can be separated primarily from Np(V), Pu(IV), Am(III) and Cm(III) ions. Further addition of ammonium carbonate leads to complex ion formation and the dissolution of actinide(IV) precipitates. However, most of the actinide(III) and (V) ions remain as precipitates under this condition. Crystalline precipitates of actinide(IV) and (VI) carbonato complex anions are formed by addition of hexamminecobalt(III), hexaureachromium(III) or hexamminechromium(III) salt to the ammonium carbonate solution containing actinide(IV) and (VI) ions.

Figure 2 shows the precipitation behavior of Pu(IV) and U(VI) carbonato complex ions. The concentration of Pu(IV) ions in the supernatant decreases gradually to a minimum ($\sim 20\mu gPu/mL$) in the range of 0.15 to 0.3M $(NH_4)_2CO_3$ when the concentration of hexamminecobalt(III) chloride is kept at 0.04M. The Pu(IV) concentration increases, then, with increasing concentration of am-

0097-6156/81/0161-0255$05.00/0

Table I Precipitates obtained

Actinide	Precipitant cation	Precipitate	Solubility in water at $20\pm3°C$ (mgAct./100gH$_2$O)
Carbonate System			
Th(IV), Pu(IV)	$[Co(NH_3)_6]^{3+}$	$[Co(NH_3)_6]_2[Act(CO_3)_5]\cdot nH_2O$	1.0
Th(IV)	$[Cr(NH_3)_6]^{3+}$	$[Cr(NH_3)_6]_2[Act(CO_3)_5]\cdot nH_2O$	1.3
Th(IV), Pu(IV)	$[Cr(CON_2H_4)_6]^{3+}$	$[Cr(CON_2H_4)_6]_2[Act(CO_3)_5]\cdot nH_2O$	0.01
U(VI)	$[Co(NH_3)_6]^{3+}$	$[Co(NH_3)_6]_4[ActO_2(CO_3)_3]_3\cdot nH_2O$	1.1
U(VI), Pu(VI), Am(VI)	$[Co(NH_3)_6]^{3+}$	$\{[Co(NH_3)_6]NO_3\}_2[ActO_2(CO_3)_3]\cdot nH_2O$	1.5
U(VI)	$[Cr(NH_3)_6]^{3+}$	$\{[Cr(NH_3)_6]NO_3\}_2[ActO_2(CO_3)_3]\cdot nH_2O$	2.0
U(VI)	$[Cr(CON_2H_4)_6]^{3+}$	$[Cr(CON_2H_4)_6]_4[ActO_2(CO_3)_3]_3\cdot nH_2O$	0.05
Sulfate System			
Th(IV), Pu(IV)	$[Co(NH_3)_6]^{3+}$	$[Co(NH_3)_6]_2[Act(SO_4)_5]\cdot nH_2O$	1.5
Th(IV)	$[Cr(NH_3)_6]^{3+}$	$[Cr(NH_3)_6]_2[Act(SO_4)_5]\cdot nH_2O$	2.0
U(VI), Np(VI), Pu(VI), Am(VI)	$[Co(NH_3)_6]^{3+}$	$\{[Co(NH_3)_6]HSO_4\}_2[ActO_2(SO_4)_3]\cdot nH_2O$	10
U(VI)	$[Cr(NH_3)_6]^{3+}$	$\{[Cr(NH_3)_6]HSO_4\}_2[ActO_2(SO_4)_3]\cdot nH_2O$	10
Oxalate System			
Th(IV), Pu(IV)	$[Cr(NH_3)_6]^{3+}$	$[Cr(NH_3)_6]_2[Act(C_2O_4)_5]\cdot nH_2O$	1.5
Th(IV), U(IV)	$[Cr(CON_2H_4)_6]^{3+}$	$[Cr(CON_2H_4)_6]_2[Act(C_2O_4)_5]\cdot nH_2O$	1.3
Pu(IV)	$[Cr(CON_2H_4)_6]^{3+}$	$[Cr(CON_2H_4)_6]_4[Act(C_2O_4)_4]_3\cdot nH_2O$	4.1

Table I (continued)

Actinide	Precipitant cation	Precipitate	Solubility in water at 20±3°C (mgAct./100gH$_2$O)
Th(IV), Pu(IV)	$[Co(en)_3]^{3+}$	$[Co(en)_3]_4[Act(C_2O_4)_4]_3 \cdot nH_2O$	3.5
Th(IV)	$[Co(en)_3]^{3+}$	$[Co(en)_3]_8[Act(C_2O_4)_6]_3 \cdot nH_2O$	0.08
Th(IV)	$[Co(tn)_3]^{3+}$	$[Co(tn)_3]_4[Act(C_2O_4)_4]_3 \cdot nH_2O$	0.01
U(VI)	$[Cr(NH_3)_6]^{3+}$	$[Cr(NH_3)_6]_4[ActO_2(C_2O_4)_3]_3 \cdot nH_2O$	3.0
U(VI)	$[Cr(CON_2H_4)_6]^{3+}$	$[Cr(CON_2H_4)_6]_4[ActO_2(C_2O_4)_3]_3 \cdot nH_2O$	5.0
U(VI)	$[Co(tn)_3]^{3+}$	$[Co(tn)_3]_4[ActO_2(C_2O_4)_3]_3 \cdot nH_2O$	6.3
Peroxide System			
U(VI)	$[Co(NH_3)_6]^{3+}$	$[Co(NH_3)_6]_4[(ActO_2)_2(O_2)_4]_3 \cdot nH_2O$	0.5
U(VI)	$[Co(en)_3]^{3+}$	$[Co(en)_3]_4[(ActO_2)_2(O_2)_4]_3 \cdot nH_2O$	3.6
U(VI)	$[Co(tn)_3]^{3+}$	$[Co(tn)_3]_4[(ActO_2)_2(O_2)_4]_3 \cdot nH_2O$	5.7
Sulfite System			
U(VI)	$[Co(NH_3)_6]^{3+}$	$[Co(NH_3)_6]_4[ActO_2(SO_3)_3]_3 \cdot nH_2O$	6.8
Citrate System			
Th(IV), Pu(IV)	$[Co(NH_3)_6]^{3+}$	$[Co(NH_3)_6]_2[Act(C_6H_5O_7)_2]_3 \cdot nH_2O$	7.5
U(VI)	$[Co(NH_3)_6]^{3+}$	$[Co(NH_3)_6]_2[ActO_2(C_6H_6O_7)_2]_3 \cdot nH_2O$	0.2
Malate System			
U(VI)	$[Co(NH_3)_6]^{3+}$	$[Co(NH_3)_6]_2[ActO_2(C_4H_4O_5)_2]_3 \cdot nH_2O$	0.9
Ethylenediaminetetraacetate System			
U(VI)	$[Co(NH_3)_6]^{3+}$	$[Co(NH_3)_6]_2[ActO_2edta]_3 \cdot nH_2O$	0.2

Figure 1.   A photograph of $[Co(en)_3]_8[Th(C_2O_4)_6]_3 \cdot nH_2O$ ($\times 3000$)

Figure 2.   Precipitation behavior of Pu-(IV) and U(VI) carbonato complex ions with hexaminecobalt(III) chloride; Pu taken, 1 mg; U taken, 1.06 mg; total volume, 1 mL: 1, Pu(IV) precipitation at 0.04M $[Co(NH_3)_6]Cl_3$; 2, Pu(IV) precipitation at 0.17M $(NH_4)_2CO_3$; 3, U(VI) precipitation at 0.04M $[Co(NH_3)_6]Cl_3$ and 0.1M NH_4NO_3; 4, U(VI) precipitation at 0.2M $(NH_4)_2CO_3$ and 0.1M NH_4NO_3

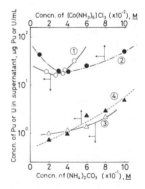

monium carbonate. In $0.17\underline{M}$ $(NH_4)_2CO_3$ solution, the concentration of Pu(IV) ions in the supernatant decreases to a minimum in the range of 0.01 to $0.04\underline{M}$ $[Co(NH_3)_6]Cl_3$. After passing through a minimum concentration it increases with increasing concentration of hexamminecobalt(III) chloride.

The concentration of U(VI) ions in the supernatant increases with increasing concentrations of both ammonium carbonate and hexamminecobalt(III) chloride above $0.2\underline{M}$ $(NH_4)_2CO_3$ and $0.02\underline{M}$ $[Co(NH_3)_6]Cl_3$. However, the concentration of U(VI) ions in the supernatant is more than 10 times lower than that of Pu(IV) ions . In this system, therefore, Pu(IV) and U(VI) ions can be separated from Am(III), Cm(III) and Np(V) ions,and 98% of the Pu(IV) or 99% of the U(VI) of the initial concentration(~1mgAct./mL) can be recovered from the solution. Furthermore, as is well known,in ammonium carbonate solution alkali earth and transition metal elements precipitate forming carbonates or hydroxides and can thus be separated from actinide(IV) and (VI) ions.

Actinide(III) precipitates dissolve to a limited extent in solution containing $1\underline{M}$ (or more) $(NH_4)_2CO_3$. However, the actinide(III) carbonato complex ion precipitates very slowly by adding hexamminecobalt(III) chloride and the yield of precipitation is not high. Separations of actinide(IV) and (VI) ions from actinide(III) and (V) ions are thus achieved by taking advantage of their different solubilities in ammonium carbonate solution. Hexamminecobalt(III) salt is used as a precipitant to recover U(VI), Pu(IV) and Am(VI) ions from ammonium carbonate solution.

Hexamminechromium(III) or hexaureachromium(III) salts may play the role of a precipitant in the place of hexamminecobalt-(III) salt. Although the yield of precipitate thus obtained is generally high in comparison with that of the corresponding hexamminecobalt(III) salt, neither of the chromium(III) complex salts is stable in the carbonate solution. Therefore, they are not recommended for the separation and recovery of actinide(IV) and (VI) ions.

Figure 3 shows a possible separation and recovery scheme of actinides based on the precipitation behavior of individual elements. A part of this procedure could be good enough if the starting solution contains simpler constituents.

Sulfate System

From dilute ammonium sulfate(pH 1-3) or sulfuric acid solutions containing actinide ions, only actinide(IV) and (VI) ions form crystalline precipitates immediately upon addition of hexamminecobalt(III) salt. Here, actinide(III) ion (e.g. Am(III) and Cm(III)) are left in the supernatant solution together with actinide(V) ion such as Np(V).

Figure 4 shows the precipitation behavior of Pu(IV) and Pu(VI) ions in ammonium sulfate solution by using hexamminecobalt(III) salt as a precipitant. The concentrations of Pu(IV) and

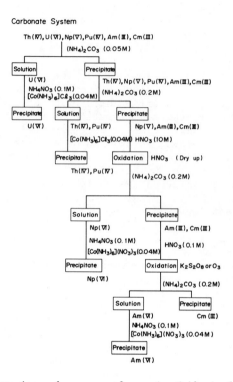

Figure 3.   Separation and recovery scheme of actinides in the carbonate system

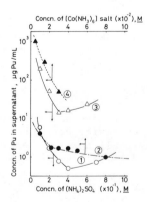

Figure 4.   Precipitation behavior of Pu(IV) and Pu(VI) sulfato complex ions with hexamminecobalt(III) salt; Pu taken, 1 mg; total volume, 1 mL: 1, Pu(IV) precipitation at 0.04M [Co(NH₃)₆]Cl₃; 2, Pu(IV) precipitation at 0.2M (NH₄)₂SO₄; 3, Pu(VI) precipitation at 0.03M [Co-(NH₃)₆](NO₃)₃; 4, Pu(VI) precipitation at 0.2M (NH₄)₂SO₄

Pu(VI) ions in the supernatant decrease rapidly to a minimum
. They increase then gradually with increasing concentration of
ammonium sulfate or hexamminecobalt(III) salt. The precipitation
behavior of Pu(VI) ion is similar to that of Pu(IV) ion. However
, the concentration of Pu(VI) ions in the supernatant is about
30 times as high as that of Pu(IV) ions when all other conditions are kept the same. In the sulfate system, Pu(IV) and U(VI)
ions can be separated from Np(V),Am(III) and Cm(III) ions. Similarly Am(VI) ion can be separated from Cm(III) ion. Thus,at initial concentration as low as 1 mg/mL more than 99% of U(VI), Pu-
(IV) and Am(VI) ions can be recovered from the solution. In the
sulfate system, hexamminecobalt(III) salt is a separating agent
as well as a precipitant of actinide(IV) and (VI) ions.

In 1M (or greater) ammonium sulfate solutions,actinide(III)
ions precipitate as double sulfates. Simultaneously hexamminco-
balt(III) sulfate also precipitates. Accordingly, attention has
to be paid to the adjustment of the ammonium sulfate concentration.

Although hexamminechromium(III) salt can also be used as a
precipitant, it is not sufficiently stable in a solution containing strong oxidants. This disadvantage makes unfeasible the
precipitation of Pu(VI) and Am(VI) ions with hexamminechromium-
(III). Figure 5 shows a possible procedure for the separation
and recovery of actinides in the sulfate system.

Oxalate System

When ammonium oxalate or potassium oxalate is slowly added
to a solution containing actinides, actinide(IV) and (VI) ions
form soluble complex ions, whereas actinide(III) ions precipitate as oxalates. Actinide(IV) and (VI) ions present in ammonium
or potassium oxalate solutions precipitate by addition of co-
balt(III) or chromium(III) complex salts.

Figure 6 shows the precipitation behavior of Th(IV) and Pu-
(IV) oxalato complex ions with hexaureachromium(III) chloride.
The concentrations of Th(IV) or Pu(IV) ions in the supernatant
depend greatly on the concentration of hexaureachromium(III) chloride and decrease rapidly with increasing concentrations of
hexaureachromium(III) chloride. On the other hand,concentrations
of ammonium oxalate in the range of 0.04 to 0.1M have little effect on this precipitation reaction. Although the precipitation
behavior of Pu(IV) ion is similar to that of Th(IV) ion,the concentration of Pu(IV) ions in the supernatant is always over 100
times higher than that of Th(IV) ions.

In oxalate media, separation of actinide(III) ions from ac-
tinide(IV) and (VI) ions is accomplished on the basis of differences of their solubilities. Hexaureachromium(III) salt may be
recommended as a precipitant for the recovery of actinide(IV)
and (VI) ions from oxalate solution. Certainly, Am(III) and Cm-
(III) ions can be effectively separated from Th(IV), U(VI) and
Pu(IV) ions.

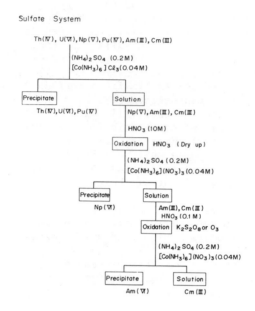

Figure 5.   Separation and recovery scheme of actinides in the sulfate system

Figure 6. Precipitation behavior of Th(IV) and Pu(IV) oxalato complex ions with hexaureachromium(III) chloride; Th taken, 4.41 mg; Pu taken, 0.88 mg; total volume, 5 mL for Th and 1 mL for Pu: 1, Th(IV) precipitation at 0.1M [Cr-(CON₂H₄)₆]Cl₃; 2, Th(IV) precipitation at 0.1M (NH₄)₂C₂O₄; 3, Pu(IV) precipitation at 0.1M [Cr(CON₂O₄)₆]Cl₃; 4, Pu(IV) precipitation at 0.1M (NH₄)₂C₂O₄

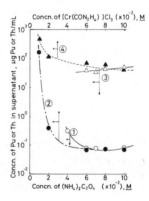

Hexamminecobalt(III) salt cannot be used as a precipitant in the oxalato complex precipitation system because it precipitates as hexamminecobalt(III) oxalate. Besides the hexaureachromium(III) salt, hexamminechromium(III), tris(ethylenediamine)cobalt(III) or tris(trimethylenediamine)cobalt(III) salts can be used as precipitants. Hexamminechromium(III) and tris(ethylenediamine)cobalt(III) salts form precipitates with actinide(IV) or (VI) oxalato complex ions, whereas tris(trimethylenediamine)cobalt(III) salt forms precipitates with Th(IV) or U(VI) oxalato complex ions leaving Pu(IV) ion in the supernatant solution. Therefore, this reagent plays the role of both a separating agent and a precipitant and is applicable for the separation of Pu(IV) ion from Th(IV) or U(VI) ion.

The oxalate system is not useful for the separation of Cm-(III) ion from Am(VI) ion because Am(VI) ion is reduced by oxalate ion and its oxalato complex precipitate like that of U(VI) ion with cobalt(III) complex ion cannot be obtained.

Peroxide System

Actinide(V) and (VI) ions form soluble complex ions with peroxide ion in slightly alkaline medium, whereas actinide(III) and (IV) ions precipitate as hydroxides. Actinide(VI) ions in slightly alkaline hydrogen peroxide solution precipitate upon addition of cobalt(III) complex salts. Figure 7 shows the precipitation behavior of U(VI) peroxo complex ion with the following kinds of cobalt(III) complex salts:
$[Co(NH_3)_6]Cl_3 > [Co(en)_3]Cl_3 > [Co(tn)_3]Cl_3$.
Here, the precipitation yield decreases with above sequence. With hexamminecobalt(III) salt, over 99% of U(VI) ion precipitates and is recovered from the solution. This complex salt is found to be a very useful precipitant for U(VI) peroxo complex ion.

The difficulties associated with this system are the decomposition of hydrogen peroxide and the resultant evolution of large amounts of gas which make the separation of the solid precipitates from liquid solution difficult. In addition, Np(VI), Pu(VI) and Am(VI) ions are reduced to Np(V), Pu(IV) or Am(III) ion by hydrogen peroxide, and their precipitates with cobalt(III) complex ion cannot be obtained. Consequently, the peroxide system is very selective for the separation and the recovery of U(VI) ion.

When hexaureachromium(III) or hexamminechromium(III) salt is used as the precipitant, the oxidation of chromium(III) ion to chromium(VI) ion proceeds with hydrogen peroxide. Accordingly, the precipitate decomposes just after being formed.

Conclusions

The advantages of the precipitation schemes discussed in this paper can be summarized as follows:

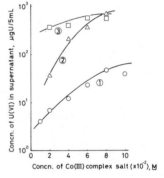

Figure 7. Precipitation behavior of U(VI) peroxo complex ion with cobalt(III) complex salt; U taken, 5.0 mg; total volume, 5 mL; H₂O₂ concn., 15%; pH 7.5–9.0; aging time, 5 min: 1, precipitation with [Co(NH₃)₆]Cl₃; 2, precipitation with [Co(en)₃]Cl₃; 3, precipitation with [Co(tn)₃]Cl₃

1. The precipitates obtained generally have a low solubility in water and thus the precipitation yield is high.
2. Double carbonate, double sulfate or double acetate has been used to separate and recover actinide so far (7,8). However,high concentrations of carbonate, sulfate or acetate are necessary for their effective precipitation. On the other hand, relatively low concentrations of sulfate or carbonate are sufficient for the precipitation of actinide in the present method. Therefore, it is favorable for treatment of waste solution after separation and recovery of actinide.
3. The rate of precipitation reaction is generally fast.The time required for the separation is reasonably short.

A possible disadvantage is that the cobalt(III) or chromium(III) complex salt is not sufficiently stable to radiolysis if the α-radioactivity of the medium is extremely high.

Literature Cited

1. Keller,C. " The Chemistry of the Transuranium Elements ";Verlag Chemie GmbH: Weinheim, 1971.
2. Koch,G.,Ed." Gmelin Handbuch der anorganischen Chemie, Transurane Teil D1: Chemie in Lösung " : Springer-Verlag: Berlin, 1975.
3. Ueno,K.; Hoshi,M. J.inorg.nucl.Chem., 1970, 32, 3817.
4. Ueno,K.; Hoshi,M. J.inorg.nucl.Chem., 1971, 33, 1765.
5. Hoshi,M.; Ueno,K. J.inorg.nucl.Chem., 1972, 34, 981.
6. Hoshi,M.; Ueno,K. J.Nucl.Sci.Technol., 1978, 15, 50.
7. Spitsyn,V.I.; Katz,J.J.,Ed. " Proceedings of the Moscow Symposium on the Chemistry of transuranium Elements "; Pergamon Press: Oxford, 1976; p.209.
8. Koch,G.,Ed." Gmelin Handbuch der anorganischen Chemie, Transurane Teil A1,II "; Springer-Verlag: Berlin, 1974.

RECEIVED December 12, 1980.

Californium-252 Encapsulation at the Savannah River Laboratory

A. R. BOULOGNE

E. I. du Pont de Nemours & Company, Savannah River Laboratory, Aiken, SC 29808

More than 1 g of the neutron-emitting isotope ^{252}Cf has been encapsulated at the Savannah River Laboratoy (SRL) for worldwide medical, industrial, and research uses. Nearly 3,000 sources for medical use (2.08 mg), and over 380 packages for industrial and research purposes have been made (1093 mg). ^{252}Cf sources and sales packages must satisfy criteria for Special Form Radioactive Material (1). Therefore, SRL performs capsule integrity and quality assurance tests on the packages and sources they have developed and produced to ensure that these criteria have been met. Bulk sales packages have been prepared for the U.S. Department of Energy (USDOE) sales program since 1971, and doubly encapsulated sources have been prepared for USDOE's market evaluation program since 1968. Encapsulation is performed in special neutron-shielded containment facilities at SRL (2,3). SRL is continually looking for ways to improve source and shipping designs and processes.

^{252}Cf is encapsulated and shipped in eight standard Special Form capsules and packages to meet the needs of the different medical, industrial, and research applications used. The isotope is available as californium oxide, californium-palladium cermet wire or pellets, and, in the case of medical therapy sources, californium-palladium cermet sheathed in platinum-iridium alloy. Capsules are available in a variety of metals and alloys (Table 1, Figures 1 through 8).

Development

SRL has developed and improved industrial, research, and medical sources since the ^{252}Cf program began. An example of the progress made can be illustrated by the history of brachytherapy sources at SRL.

TABLE 1

Standard Source Forms

Model	Figure	Description	Use
SR-Cf-XX	1	Primary capsule for "point" sources of ^{252}Cf oxide	For loan in the market evaluation program
SR-Cf-1X	2	Primary capsule for "line" sources of ^{252}Cf cermet wire or point sources of cermet pellets	For loan in the market evaluation program
SR-Cf-100	3	Secondary capsule (may contain either SR-Cf-1X or SR-Cf-XX primary capsule)	For loan in the market evaluation program
SR-Cf-1000	4	^{252}Cf shipping capsule assembly	For all shipments of ^{252}Cf purchased by encapsulators and users (may be in the form of oxide, cermet wire, or cermet pellets)
ALC-X	5	Radiotherapy after-loading cell for interstitial implantation	For loan in the market evaluation program
SALC-X	6	Short afterloading cell for interstitial implantation	For loan in the medical evaluation program
AT-X	7	Radiotherapy applicator tube for intracavitary implantation	For loan in the medical evaluation program
SEEDS (ALC-P4C)	8	Radiotherapy seed assemblies for interstitial implantation	For loan in the medical evaluation program

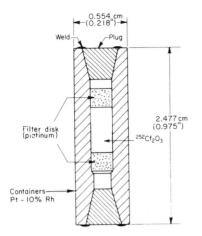

Figure 1. Primary capsule for "point" sources of ^{252}Cf oxide (SR–Cf–XX)

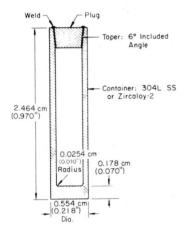

Figure 2. Primary capsule for "line" sources of ^{252}Cf cermet wire or "point" sources of cermet pellets (SR–Cf–1X)

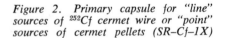
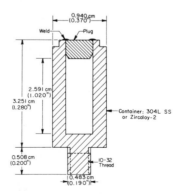

Figure 3. Secondary capsule (may contain either SR–Cf–1X or SR–Cf–XX primary capsule) (SR–Cf–100)

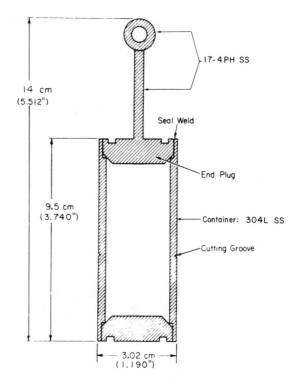

Figure 4. ²⁵²Cf *shipping capsule assembly (SR–Cf–1000)*

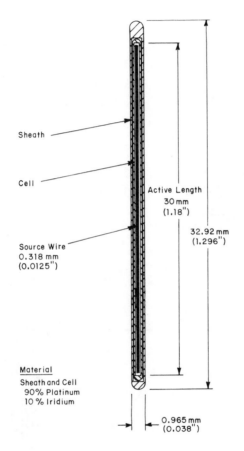

Figure 5. Radiotherapy afterloading cell for interstitial implantation (ALC–X)

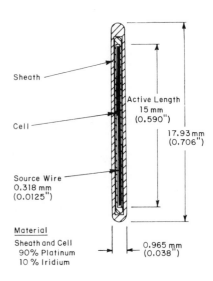

Figure 6. Short afterloading cell for interstitial implantation (SALC–X)

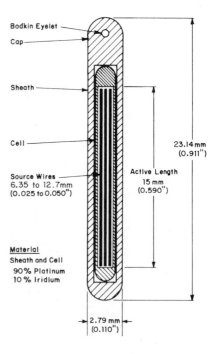

Figure 7. Radiotherapy applicator tube for intracavitary implantation (AT–X)

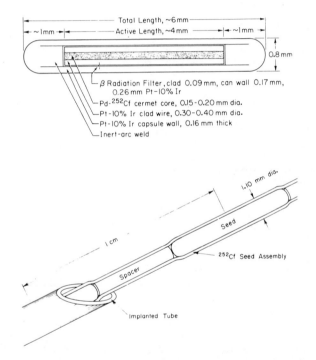

Total Length, ~6mm
Active Length, ~4mm
~1mm
~1mm
0.8 mm

β Radiation Filter, clad 0.09 mm, can wall 0.17 mm,
0.26 mm Pt–10% Ir
Pd–^{252}Cf cermet core, 0.15–0.20 mm dia.
Pt–10% Ir clad wire, 0.30–0.40 mm dia.
Pt–10% Ir capsule wall, 0.16 mm thick
Inert-arc weld

1.10 mm dia.
Seed
1 cm
^{252}Cf Seed Assembly
Spacer
Implanted Tube

Figure 8. Radiotherapy seed assemblies for interstitial implantation (Seeds) (ALC–P4C)

Brachytherapy Sources. The first ^{252}Cf sources for radiotherapy research were prepared at SRL about fourteen years ago
(4,5). Initially, these sources resembled the classical radium
needles familiar in clinical radiotherapy. Eventually, afterloading cells and applicator tubes were supplied to medical
evaluators, and all medical sources were improved by the use of
californium-palladium cermet wire sheathed in Pt - 10% Ir alloy
(6,7). The most recent designs for medical sources produced in
quantity for therapy research are line sources and point sources
containing a range of ^{252}Cf from less than 1 μg to 200 μg (Figures
9 and 10).

The purpose of the development work at SRL is to prepare
physically small, yet intense ^{252}Cf sources for remote afterloading brachytherapy research. Remote afterloading is used by
many hospitals to avoid radiation exposure to medical and
hospital personnel.

Typically, a remote afterloader is a shielded cell or treatment room. Long cables pass through the wall of the room from an
operating station outside the room to a patient treatment position
inside the room. Radioactive sources are attached to the cables
at the patient treatment position. With the patient in the treatment room and the shielded room door closed, the sources are
advanced to the treatment position in previously installed appli-
cators by means of the cables and a mechanism in the control
system outside the treatment room wall. Our goal was to adapt
sources containing up to 200 μg of ^{252}Cf to use in these remote
afterloaders.

At least five designs of remote afterloaders are available:
the "Curietron" in France, the "Cervitron-II" in Switzerland, the
"Cathetron" in England, the "Hicesitron" in the United States, and
the "Brachytron" in Canada. The afterloading sources described in
this paper were designed specifically for the "Brachytron" manufactured by the Atomic Energy of Canada, Ltd., and for a remote
afterloader manufactured by Toshiba Electric Co., Ltd., Japan.

Hardware for the "Brachytron" was initially designed for use
with ^{60}Co sources; straight or curved catheters are available.
The "stiff" length of the source is defined by the minimum radius
through which the source must travel. Available catheters will
accept sources whose "stiff" length is between 15.2 mm and 24.4 mm.
We have prepared 200 μg ^{252}Cf sources with the shortest possible
"stiff" length and the greatest possible integrity.

These sources were prepared by a modified chemical plating
technique similar to that used to prepare palladium-californium
oxide cermet for industrial applications (6). Design of the

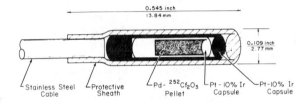

Figure 9.   Cf packaging facility

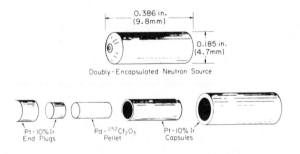

Figure 10.   Cf packaging section

doubly encapsulated source with its protective sheath is shown in Figure 9. Existing "Brachytrons" can accommodate this source form. This capsule system will withstand internal gas pressure from helium due to alpha decay and fission gases from a 200 µg source after ten years if the source is subjected to a maximum temperature of 800°C (the assumed temperature of an accidental fire). Under these conditions, the safety factor is 9.

Facilities

Sales Packages and Industrial Sources. All industrial encapsulations and the preparation of cermet pellets for medical sources are performed in a facility that limits radiation exposure rates to less than 1 mrem/h in normally occupied areas. Maximum amounts of up to 100 mg of ^{252}Cf may be handled in this shielded facility. Figures 11 and 12 show the cell complex, which is surrounded by a maximum of 1.22 m of gamma and neutron shielding. Because the chemical and encapsulation procedures require micro-techniques, in-cell equipment has been designed for remote handling, assembly, and examination of microliter quantities of solutions and capsule components ranging in size from 0.317-cm-long cylinders (porous disks) to 15.74-cm-long outer capsules (sales packages). This remotely operated equipment includes an ion column leaching apparatus, an automatic pipettor, special dies and fixtures for hydraulic presses, a programmable tungsten arc, inert-gas shielded welder, an ultrasonic decontamination system, electronic assay systems (fission counters and BF_3 counters), and a portable in-cell fast neutron monitor. Small samples can be transferred from three of the five operating positions to an adjacent laboratory by a pneumatic transfer system.

Medical Sources. All medical encapsulations are done in a facility which consists of a series of five interconnected stainless steel boxes which provide primary containment of process equipment and materials. Interconnections include transfer ports, drop-through tubes, and ventilation ducts. Each containment enclosure has a floor area of 1.52 m x 0.91 m; the floor level is 0.76 m above the building floor. A pair of master-slave manipulators serve each containment box. A closed circuit television system and a telescope are provided for close-up viewing of in-cell operations and can be moved from cell to cell as needed. General arrangement of the facility is shown in Figures 13 and 14.

Two glove boxes at the rear of the facility are for material entry and exit operations. Two additional glove boxes provide containment for the access tubes that are used for removal of swipes used to assess effectiveness of decontamination procedures.

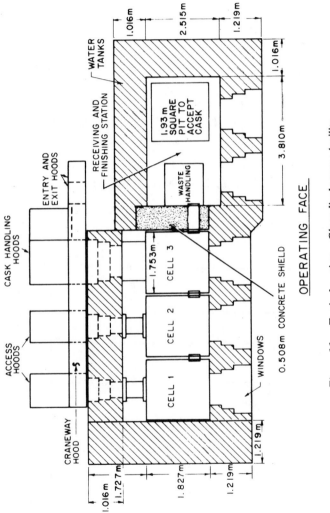

Figure 11. Typical section—Cf medical source facility

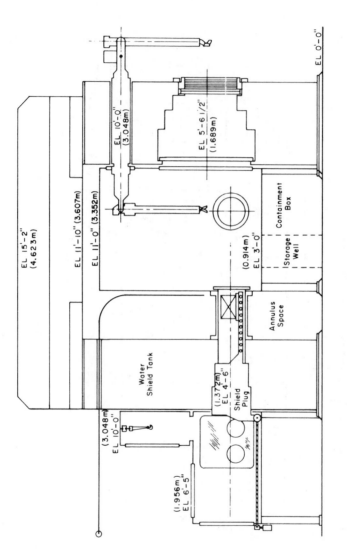

Figure 12.   Plan view—Cf medical source facility

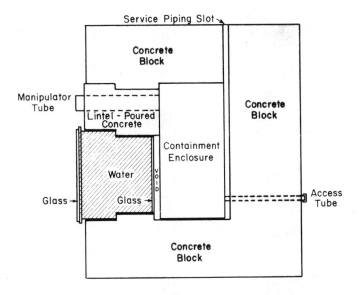

Figure 13. ^{252}Cf *"Brachytron" source*

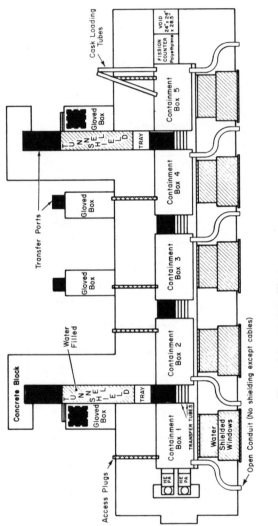

Figure 14.   Toshiba ²⁵²Cf afterloading source

The containment cells are shielded on the front, back, ends, and top by concrete blocks. There is no shielding between containment enclosures. Wall thickness of the blocks is approximately 0.99 m. Shield windows are installed in the front face of the containment enclosures. The inner plate is 2.54 cm radiation-stabilized glass; the outer plate is 2.29 cm Pyrex borosilicate glass. Space between these plates is filled with 0.91 m water, which circulates continuously through a filtration-clarification system. Manipulators are mounted in concrete filled lintels. Concrete blocks, totaling 0.77 m thickness, shield the top of the facility.

Normal production requires Containment Boxes 1 and 2 to contain approximately 400 µg of ^{252}Cf as californium-palladium cermet. While operating in this mode, general area dose rates are about 1 mrem/h where personnel usually operate in front of the boxes. Dose rates behind the cells in normally unoccupied areas range from 4 to 6 mrem/h (3).

Medical sources are fabricated with remotely operated, specially designed machines (8). The fabrication process involves production of Pt - 10% Ir-clad wire with a californium oxide-palladium cermet core. The wire is swaged and drawn to size, cut to length, and welded in a Pt - 10% Ir capsule. Nominally, medical sources contain from 0.3 µg ^{252}Cf in an individual seed (Figure 8) to 30 µg ^{252}Cf in an applicator tube (Figure 7).

Feed to the medical source facility is a composite billet prepared in the sales and industrial source packaging facility. The billet contains a californium oxide-palladium core gold-brazed in a Pt - 10% Ir container (0.76-cm diameter x 3.56 cm long). The core is prepared by deposition of palladium on a fine precipitate of californium oxalate in an aqueous system (6).

The clad wire is made by swaging and drawing. This wire provides a source form that can be subdivided with a minimum of contamination. The swaging operation requires five reductions and one anneal at 1100°C to reduce the billet from 7.6 mm to 2.5 mm in diameter. The drawing operation requires approximately 44 passes, which reduces the cross-sectional area of the wire by 10% per draw. Four annealing cycles at 1100°C are required during the drawing operation. These operations produce a finished wire with a reasonably symmetrical cross-section (8). The swager-draw bench, a precision cutter, loading and welding equipment, and the leak testing apparatus were designed specifically for this process.

Capsule Integrity

The integrity of the design and construction of each source and sales package was demonstrated by subjecting active sources to tests simulating expected adverse service conditions as specified in "Tests for Special Form Materials" (1). These tests include:

- Impact (falls 10 m on unyielding surface)
- Percussion (1.4-kg weight, 25-mm diameter, falls 1 m)
- Heating (10 minutes at 800°C)
- Immersion (24 hours in water at room temperature)

Special tests not specified by U.S. regulatory agencies were conducted to simulate expected service conditions. Dummy secondary capsules of the SR-Cf-100 series were subjected to internal and external pressures far in excess of pressures expected under the most adverse industrial service conditions. The special tests that successfully demonstrated the integrity of source construction and the seal welds were:

- Burst tests of the circumferential welds
- Crush tests
- Hydrostatic compression tests

Similarly, special tests to demonstrate the integrity of the ^{252}Cf seed assemblies (ALC-P4C) under conditions that might conceivably be expected during medical service were:

- Sterilization
- Pinching
- Crushing
- Bending

In addition to all of the above tests, quality assurance tests during encapsulation demonstrate the integrity of each ^{252}Cf sales package and neutron source.

Quality Assurance

Californium Assay and Analyses. Quality control for the californium feedstock is accomplished by measuring the neutron emission rate of an aliquot of the starting material and by performing analyses for isotopic content and chemical purity. Neutron emission rate is measured in a fission counter (9). Isotopic content is measured by mass spectrometry and chemical purity by spark source mass spectrometry. The completed assembly is leak tested, decontaminated, and assayed before packaging and shipping.

Inspecting Capsule Components. Prior to cleaning, capsule
components are inspected for dimensional accuracy and machining
flaws.

Cleaning Capsule Components Prior to Use. Metal components
are thoroughly degreased and cleaned to remove cutting oil,
grease, fingerprints, and dirt.

Welding Control. Plugs in the inner and outer capsules for
the sales packages and industrial sources are seal-welded with an
argon-shielded or helium-shielded tungsten electrode DC arc. The
capsule is rotated under the automatically controlled arc to
produce a minimum weld penetration of 1.27 mm. Each weld bead is
visually inspected by periscope or by Questar telescope, and
imperfectly formed welds are rejected. Weld quality is controlled
by periodic metallographic examination of dummy capsules welded in
the in-cell equipment.

Capsule closures for medical sources and seeds are made with
an argon-shielded plasma DC arc. The arc is controlled to produce
a weld bead penetration equal to, or greater than, the capsule
wall thickness. Each weld bead is visually inspected by a 20X
stereoscopic microscope or by Questar telescope. Weld quality is
controlled in the same manner as with sales packages and industrial sources.

Leak Testing. Sealed capsules of sales packages and industrial sources are pressurized in 300 psi helium for 20 minutes.
Leak tests are performed on individual capsules in a helium leak
detector whose lower limit of detection is 1.0×10^{-8} standard
cubic centimeters of helium per second. All capsules must show no
detectable leak.

Leaks in medical sources and seeds are detected by a vacuum
immersion leach test. Because the internal volume of the medical
sources is so small (3.6×10^{-4} mL, in the case of the ALC-P4C
seed), the conventional helium leak test is not a valid leak test
procedure. About 45 minutes is required to pump down the system
before helium measurement is begun. If the internal volume of the
test specimen is small, trapped helium would escape before helium
assay begins. Therefore, leaks in encapsulated medical sources
are detected by measuring the alpha activity of a nitric acid
penetrant solution in which the source had been immersed. After
immersion, pressure above the liquid is decreased to 2.5 psia for
3 min before venting to atmosphere. This procedure is repeated
twice, then the sources remain in acid a minimum of 16 h at 20°C.
A 1-mL sample of the leach solution is assayed for alpha radio-
activity, and the sources are rejected if the alpha count exceeds
10 d/m above background.

Labeling. Sales packages are identified by engraved designations which are numbered serially (SR-Cf-1001, SR-Cf-1002, etc.). Industrial sources are also numbered serially in the -100 series (SR-Cf-101, SR-Cf-102, etc.). Each package and source is provided with an information sheet listing pertinent construction, test, and calibration data.

Medical sources and seeds are not identified by engraving so as not to lessen the integrity of the 0.015 mm wall. Successive groups of sources made for each contractor are identified by uniquely positioned gold bands, or other color coding as requested by the user to accommodate handling procedures. Information sheets are furnished as with sales packages and industrial sources.

Neutron sources for the Toshiba Electric Co. afterloader can be slightly larger in physical size than "Brachytron" sources and must also contain approximately 200 µg of ^{252}Cf. The source carrier will accept a doubly encapsulated source 9.8 mm long x 4.7 mm diameter (Figure 10). The two capsules are made of Pt - 10% Ir alloy and are seal-welded with an argon-shielded tungsten electrode DC arc. We have prepared eight of these sources containing nanogram amounts of ^{252}Cf and have subjected them to tests specified for the Special Form Materials (1) and to other adverse service conditions that include sterilization, pinching, crushing, and abrasion. The sources successfully passed all safety tests.

Using the same process, we prepared and shipped three nominal 200 µg ^{252}Cf sources to the Keio University School of Medicine in Tokyo, Japan for evaluation in the Toshiba Electric Co., Ltd. afterloader. Three additional sources are currently being fabricated for the same evaluator.

Storage and Shipping

Although special containment and shielding considerations apply to ^{252}Cf handling, the problems of radiation protection are straightforward (10). Sources containing more than a few micrograms of ^{252}Cf must be handled remotely and stored in hydrogenous shields rather than high density materials, such as load.

During the past fourteen years, special equipment, procedures, and carriers have been developed for safe, practical handling, storage, and shipment of encapsulated ^{252}Cf sources. Tools and fixtures were designed to attach magnetic eyelets for remote transfer of finished sources. Sources are stored in large polyethylene tanks filled with a cast matrix of water-extended polyester (WEP) (11). Boric acid for neutron absorption and ethylene glycol for freeze protection are incorporated in the aqueous phase. Figure 15 is a photograph of a 1900-L polyethylene tank filled with WEP.

Figure 15.   WEP-shielded ^{252}Cf storage

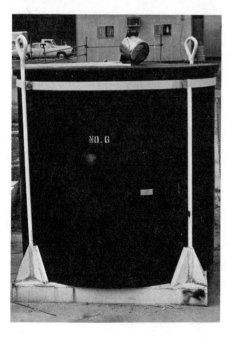

Figure 16.   SRL ^{252}Cf transfer cask

Figure 17.   SRL 4.5-ton Cf shipping cask

The central storage well is a 10.16-cm pipe surrounded by 5.08 cm of lead for primary gamma shielding. The radiation dose rate at ∿1 m from this container with 8 sources totaling 34.6 mg ^{252}Cf in the center well was 55 mrem/h (42 mR/h gamma + 13 mrem/h neutron radiation).

Fourteen years ago, a few shipping casks with sufficient shielding to accommodate 200 μg of ^{252}Cf were available; however, none existed for handling milligram quantities. Since that time, shipping casks have been constructed at U.S. Department of Energy sites for transporting Type A and Type B quantities of ^{252}Cf and at commercial cask fabricators for transporting Type A quantities (12). These casks satisfy Department of Energy, Nuclear Regulatory Commission, and Department of Transportation specifications.

Type A packaging means packaging which is designed in accordance with the general packaging requirements and is adequate to prevent the loss or dispersal of the radioactive contents and to retain the efficiency of its radiation shielding properties if the package is subject to normal transportation.

Type B packaging means packaging which meets the standards for Type A packaging, and, in addition, meets the standards for hypothetical accident conditions for transportation (1).

Maximum quantity of ^{252}Cf in Special form packaging permitted for Type A shipment is 2.0 Ci (3.73 mg). Figure 16 shows a typical Type A cask permitted for shipment of 2.0 Ci ^{252}Cf in a Special Form package. This SRL ^{252}Cf Transfer Cask uses Benelex 401 for shielding, is 1.22 m wide x 1.32 m high, octagonal in shape, and weighs 2386 kg.

Figure 17 shows the SRL 4.5-ton Californium Shipping Cask for shipment of Type B quantities in Special Form packaging (13). This cask is identified as USA/6642/B. It has USNRC Certificate of Compliance 6642 and International Atomic Energy Agency (IAEA) Certificate of Competent Authority USDOT USA/6642/B. The cask is a 1.52 m diameter steel sphere filled with WEP as the neutron shield. It weighs 4332 kg and can carry 46 Ci (85 mg) ^{252}Cf in Special Form packaging.

ACKNOWLEDGEMENT

The information contained in this article was developed during the course of work under contract No. DE-AC09-76SR00001 with the Department of Energy.

LITERATURE CITED

1. Code of Federal Regulations, "Transportation," Title 49, Paragraph 173.398, April 23, 1976.

2. Boulogne, A. R.; "Californium-252 Encapsulation and Shipping at SRL," USERDA Report CONF-720902, Applications of Californium-252, ANS National Topical Meeting, Austin, TX, September 11-13, 1972, (1975), 36.

3. Moyer, R. A.; Safety Analysis of the Californium Medical Source Facility, USAEC Report DPSTSA-700-11, E. I. du Pont de Nemours & Co., Savannah River Laboratory, Aiken, SC, 1973.

4. Boulogne, A. R.; Evans, A. G.; "Californium-252 Neutron Sources for Medical Applications," Int. J. Appl. Rad. Iso., 1969, 20, 453.

5. Wright, C. M.; Boulogne, A. R.; Reinig, W. C.; Evans, A. G.; "Implantable Californium-252 Neutron Sources for Radiotherapy," Radiology, 1967, 89 (2), 337.

6. Mosley, W. C.; Smith, P. K.; McBeath, P. E.; "Neutron Sources of Palladium-Californium-252 Oxide Cermet Wire," USERDA Report CONF-720902, Proceedings of the American Nuclear Society Topical Meeting, September 11-13, 1972, (1975), 51-60.

7. Walker, V. W.; "Equipment and Operations for Preparing Neutron Sources for Interstitial Cancer Radiotherapy Research," Trans. Am. Nucl. Soc., 1975, 22, 730.

8. Permar, P. H.; Walker, V. W.; "Californium-252 Radiotherapy Sources for Interstitial Afterloading," USAEC Report CONF-760436, Proceedings of the Paris Symposium of the International Symposium on Californium-252 Utilization, Paris, France, April 26-28, 1976, Volume II, Session II, 145-159.

9. Herold, T. R.; "Electronic Assay of ^{252}Cf Neutron Sources," Nuclear Technology, 1972, 14, 269.

10. Wright, C. N.; "Radiation Protection for Safe Handling of Californium-252 Sources," Health Physics, 1968, 15, 446.

11. Oliver, G. D., Jr.; Moore, E. B.; "Neutron Shielding Qualities of Water-Extended Polyesters," Health Physics, 1970, 19, 578.

12. Californium-252 Progress, Numbers 1 through 22, Savannah River Operations Office, U.S. Department of Energy, Aiken, SC, 1970–1978.

13. Whatley, V., Jr.; Mahoney, D. J.; Livingston, J. T.; Safety Analysis Report - Packages, SRL 4.5-Ton Californium-Shipping Cask, USAEC Report DPSPU-74-124-6, Rev. 1, E. I. du Pont de Nemours & Co., Savannah River Laboratory, Aiken, SC, March 1976.

RECEIVED February 9, 1981.

INDEX

INDEX

Jacket design by Carol Conway.
Production by Robin Giroux and V. J. DeVeaux.

Elements typeset by Service Composition Co., Baltimore, MD.
The book was printed and bound by The Maple Press Co., York, PA.

DATE DUE

OCT 29 1996			
NOV 26 1996			

DEMCO 38-297